Table of Contents

Introduction

Everyone wants to spend less his/her free time preparing breakfast, dinner, but spend your own time with family, relatives and friends. This time is very valuable and because of it, people were finding something new and helpful. Now the resolution was found. If you have purchased the Air Fryer already and have it at your place, you will get more free time comparing with the previous times, and you can spend it as you wish, while breakfast or dinner or even the scrumptious dessert is preparing. So, which benefits does the Air Fryer have for you? How to improve your life for better? What should be done for it?

Today, the modern women don't want to work during the entire day at kitchen, but they have the great desire to cook different meals quickly for their families. So, is it imaginable to do? This issue can be basically solved with the Air Fryer. And life of all people will be easier, because the solution was found. So, everyone can check it and improve his/her life.

The main purpose, which Air Fryer has, is unassuming and easy– you place the ingredients and the Air Fryer cooks it and you will hear the sound signal when food has been already prepared. After that you can take scrumptious meal out of Air Fryer and enjoy with your family. It is possible to make the huge list of different meals in the Air Fryer without a lot of efforts for people. There are different methods of cooking food in the Air Fryer. You just should choose: breakfast or lunch, main dish or dessert and so on. Just place products, press the button and do what you want. It is really appropriate. Even the cheap Air Fryer can prepare an enormous number of different food and it will not take a lot of place on your kitchen. There are only pluses in purchasing of this useful product. The process of cooking food is controlled by microchip. You can set the needed time, temperature and even change it when food is preparing. This process is entirely automatic and you can easily prepare any type of food. It is really great, that you should not do anything – this process of preparing is controlled by the robot and the system will do everything for you. It is really simple and convenient. A lot of Air Fryer models have some unusual types for preparing the food, but it depends on the model and on the price. Today you can purchase any Air Fryer you want. It is possible to find the different information on the Internet, compare the prices and read the feedbacks of the people that bought the Air Fryer and used it. After that, you can make the decision and purchase the needed model for you. It is significant to know all information about the Air Fryer before you buy it, because it is the thing for the long period of using. You can ask the consultants in the shop, check the quality of the Air Fryer, open it and taste. It is better to spend your time for the choosing of model and purchasing and after that just enjoy with the Air Fryer cooked food. You will spend a few minutes more, but later, you will have the great helper for yourself on the kitchen.

It is possible to prepare the different sorts of meal in the Air Fryer. You can cook different snacks, main dishes, meat, soups and desserts. Everything will be prepared quickly, flavorsome and the main thing is the fact that it is vigorous food. The vitamins, which are in the products will be there when you cook the food. Because of this fact, you can be sure that you are eating the healthy food with the huge number of minerals and vitamins. A lot of people take the Air Fryer for their families exactly because of this fact. It is very important to eat the vigorous food and the Air Fryer can help you with it. You can even prepare the easiest thing in the Air Fryer - eggs. Really, you just need to place them on the special grill and eggs will be ready in a few minutes. Also, you can cook sausages in the Air Fryer. You just need to place meat in the Air Fryer and set up the needed temperature. Later you will get the soft meat which will melt in your mouth and you will enjoy with this delicious and tasty meat.

The big plus of the Air Fryer is the fact that it is possible to put the products in the evening and it will cook breakfast in the morning at exact time you need. It is a very useful thing for people that do not have a lot of time in the morning for cooking something to it. The Air Fryer will save your time and you will have more free time for yourself in the morning.

The Air Fryer has the function of the preheating the food. It means that when you prepare the food, it will be warm during 12 hours. It is ideal option to put the products in the Air Fryer late in the evening and go to sleep. You will get a lot of energy in the morning after eating breakfast, which was prepared in the Air Fryer during the whole night. The food will be warm and appetizing. All what you need to do - just to clean the Air Fryer. Also, this option is great if you prepare dinner for your children, but they are late and you should not cook it again. Some people are sure in the fact that the Air Fryer needs a lot of efforts, but it is easy to clean it. You just need to clean the bowl, dry and set it back. It does not take a lot of place. Also, it can replace the pan, bowl and so on. After using, you can just clean and dry it, put back in the box and put somewhere and it will not disturb you. This fact is great if your kitchen is not too big and you do not have a lot of space. To have more free space, you can save the Air Fryer even under dinner table. As you can see, it is convenient and simply.

You can cook everything fast with the help of the Air Fryer and it does not need you to stay near it and check. Just set the desired temperature and time and you are free. As soon as the food is ready, the Air Fryer will give you the signal. It is significant to get breakfast or dinner in a few minutes, but not to wait for it, for example, 20 minutes or more. Today, every minute is precious because it is better to do anything else than to spend your time and cooking. You even should not use oil for cooking the food. It means that the food, which is roasted in the Air Fryer will not be very fat as the food, which is prepared in the usual pan, for example. It will be better for your health because you will need only sprinkle the frying basket with oil in some cases. If you cook meat, for example, you can omit this step. If you have unexpected guests, it is not difficult to prepare something appetizing and great for them. It will take a few minutes of your time if you cook something easy and modest. At that time, you can speak with them and you will not need to go to the kitchen for checking the food every minute. The Air Fryer is the robot, which can help you to cook the food and it is a significant thing in the house.

The Air Fryer is safe. It means that your children can play near this device and it will be ok. You can take the Air Fryer even if you have the vacation or you are travelling. It will not take a lot of place in your car, but you will forget about cooking the food there and will enjoy with your rest. You can spend your time with your relatives and the Air Fryer will cook the food instead of you.

You should not have some unusual knowledge for preparing the food in the Air Fryer. Even the child can make something. You just need to place the products and choose the temperature. Also, every Air Fryer has the special instruction, where you can find the information about using it in your everyday life. Firstly, it can be unusual for you, but later, when you know all types of cooking in the Air Fryer, all temperatures, it will be uncomplicated for you to cook something. Even if you are new at cooking, you should not worry about it. The detailed explanation in the instruction of the Air Fryer will help you a lot and you can prepare the most complicated meals on your kitchen with the help of it. Some people can think that the Air Fryer is difficult, but they just need to have more time for using it. They just used to cook in the traditional way and do not want to change anything in their life because they are afraid of the changes. However, if you purchase the Air Fryer, your life will be changed only for better. As you can see, there are only benefits in having the Air Fryer at home. After that purchasing of the Air Fryer you will get the useful thing in your home, which will cook the food instead of you and the result will be great and appetizing. Do not miss your chance! It is really needed thing for your kitchen and you should purchase it.

Breakfast

Scrumptious Potatoes

It is not difficult and flavorsome for breakfast. The delicious meal with crispy skin.

Prep time:5 minutes

Cooking time:30 minutes

Servings:4

Ingredients:

- 9 oz of potatoes
- ½ teaspoon of salt
- 1 teaspoon of pepper
- 1 teaspoon of cumin
- ½ teaspoon of oregano
- ½ teaspoon of chili pepper
- ½ tablespoon of dry herbs

Directions:

1. Rinse and clean potatoes.
2. Slice them in the pieces.
3. Take the container and blend salt, pepper, cumin, oregano, chili pepper and dry herbs there.
4. Then rub potatoes with these seasonings.
5. Preheat the Air Fryer to 320F.
6. Sprinkle the Air Fryer with oil.
7. Place potatoes and cook for 25 minutes at 320F.
8. Then shake it and cook for 5 minutes at 350F.
9. Serve hot with sauce or vegetables.
10. Eat appetizing meal.

Nutrition:

- Calories: 192
- Fat: 9,8g
- Carbohydrates: 23,4g
- Protein: 2,8g

Grilled and Appetizing Tomatoes

This meal is delectable and mouthwatering. Prepare these flavorsome tomatoes for breakfast.

Prep time:5 minutes

Cooking time:20 minutes

Servings:2

Ingredients:

- 2 tomatoes
- ½ teaspoon of salt
- ½ teaspoon of pepper
- ½ teaspoon of garlic
- ½ teaspoon of cumin
- ½ teaspoon of oregano
- 2 tablespoons of oil

Directions:

1. Rinse and clean tomatoes.
2. Slice them in two pieces.
3. Grease them with the tablespoon of oil.
4. Then blend salt, pepper, oregano, cumin with garlic.
5. Rub potatoes with this combination of flavors.
6. Preheat the Air Fryer to 320F for 20 minutes.
7. Cook tomatoes in the Air Fryer for 10 minutes.
8. Shake them well.
9. Cook the meal for 10 minutes.
10. Serve the hot meal. Decorate it with parsley or basil.
11. Enjoy with the meal.

Nutrition:

- Calories: 56,5
- Fat: 3,9g
- Carbohydrates: 4,23g
- Protein: 1,1g

Potato Hash

Do you want something new and delicious for breakfast? Try this potato hash. It is flavorsome and great.

Prep time:15 minutes

Cooking time:40 minutes

Servings:4

Ingredients:

- 2 eggs
- 1 teaspoon of salt
- ½ teaspoon of pepper
- ½ green pepper
- ½ teaspoon of thyme
- ½ teaspoon of paprika
- 2 tablespoons of oil
- 1 onion

Directions:

1. Sprinkle the frying basket with oil.
2. Preheat the Air Fryer for 2 minutes to 350F.
3. Cut onion in the pieces, place in the Air Fryer. Cook for 2 minutes.
4. Then chop pepper, add in the Air Fryer to onion, blend everything and cook for 5 minutes more.
5. After that wash and clean potatoes.
6. Cut them in the pieces.
7. Blend salt, pepper, thyme and paprika in the bowl.
8. Add these seasonings to potatoes and blend them.
9. Then place potatoes in the Air Fryer and cook for 30 minutes at 350F.
10. Beat eggs in the bowl.
11. Then pour potatoes with eggs, add pepper and onion and cook for 5 minutes at the same temperature.
12. Serve hot with sauces.

Nutrition:

- Calories: 213
- Fat: 8,6g
- Carbohydrates: 6,7g
- Protein: 2,9g

Sweet Polenta Bites

These sweet bites will be great for morning tea or coffee. They are easy in preparing and very delicious.

Prep time:10 minutes

Cooking time:16 minutes

Servings:4

Ingredients:

- 2 tablespoons of oil
- 1 tablespoon of coconut oil
- ¼ cup of flour
- 1 cup of cornmeal
- 2 cups of water
- 1 tablespoon of butter
- ½ teaspoon of salt

Toppings:

- Powdered sugar
- Jam
- Maple syrup

Directions:

1. Boil water with salt.
2. Then add cornmeal.
3. Mix everything well and cook.
4. Then add the tablespoon of butter.
5. Put the mixture in the fridge for 20 minutes.
6. It should not be hot.
7. Then add coconut oil to the dough.
8. Make the balls from this mixture.
9. Cover them with flour.
10. Then preheat the Air Fryer to 320F.
11. Cook them for 8 minutes.
12. Serve warm with different toppings.

Nutrition:

- Calories: 360
- Fat: 9,8g
- Carbohydrates: 8,7g
- Protein: 3,5g

Scrumptious and Crispy Potatoes

This tempting recipe is for you. Cook and try. The meal is really flavorsome.

Prep time:30 minutes

Cooking time:35 minutes

Servings:4

Ingredients:

- 1 teaspoon of onion powder
- 1 teaspoon of pepper
- 1 teaspoon of garlic powder
- 2 teaspoons of salt
- 2 tablespoons of oil
- 1 red pepper
- 1 onion
- 6 big potatoes

Directions:

1. Rinse and clean potatoes.
2. Cut and place potatoes in the container.
3. Rub with pepper, garlic powder, salt
4. Sprinkle the Air Fryer with oil.
5. After that place potatoes in the Air Fryer.
6. Cook for 20 minutes at 350F.
7. Then slice onion in the pieces.
8. Cut pepper in the small parts and blend them with onion.
9. Add to potatoes and cook at the same temperature for 10 minutes.
10. Then blend and cook for 5 minutes.
11. Serve hot food with parsley or ketchup.
12. Enjoy with the great meal.

Nutrition:

- Calories: 280
- Fat: 9,3g
- Carbohydrates: 8,2g
- Protein: 3,1g

Toasted Bagels

It will be your favorite meal for breakfast. There is nothing difficult in preparing and after eating this delicious breakfast you will have a lot of energy.

Prep time:2 minutes

Cooking time:6 minutes

Servings:2

Ingredients:

- 2 bagels
- 1 tablespoon of jam
- 1 tablespoon of chocolate
- 1 teaspoon of icing sugar
- 1 tablespoon of butter

Directions:

1. Sprinkle the Air Fryer with oil.
2. Preheat it to 350F.
3. Cut bagels in 2 pieces.
4. Put them in the Air Fryer and cook for 3 minutes.
5. Then cover bagels with butter and cook for 3 minutes more.
6. Take out of Air Fryer and put on the plate.
7. They should not be very hot. It is possible to put them in the fridge.
8. Cover with the icing sugar, jam and melted chocolate.
9. Eat with tea, coffee, juice or milk.
10. This meal is really tasty.

Nutrition:

- Calories: 270
- Fat: 7,5g
- Carbohydrates: 7,5g
- Protein: 1,1g

Scrumptious Cinnamon Toast

This toast is really perfect. It will take only 5 minutes for cooking. Breakfast will be appetizing.

Prep time:5 minutes

Cooking time:5 minutes

Servings:4

Ingredients:

- 1 tablespoon of vanilla
- 1 teaspoon of ground cinnamon
- ½ cup of sugar
- 1 stick salted butter
- 12 pieces of bread
- 2 tablespoon of oil

Directions:

1. Take the container and put butter there.
2. Then add sugar, cinnamon and vanilla.
3. Blend the components.
4. Take the pieces of bread and cover one side with this mixture.
5. Sprinkle the Air Fryer basket with oil.
6. Place the pieces of bread with the combination in the Air Fryer and cook for 5 minutes at 390F.
7. Place on the dish and cut in 2 pieces.
8. Serve warm with different toppings or ice cream.
9. Enjoy with this perfect meal.

Nutrition:

- Calories: 157
- Fat: 6,2g
- Carbohydrates: 2,3g
- Protein: 1,2g

Delectable and Peppery Frittata

It is something wonderful and flavorsome. Then cook this appetizing frittata.

Prep time:5 minutes

Cooking time:5 minutes

Servings:4

Ingredients:

- ½ teaspoon of salt
- ½ teaspoon of pepper
- 1 tablespoon of oil
- 1 tablespoon of parsley
- 4 small tomatoes
- ½ of Italian sausage
- 3 eggs
- ½ teaspoon of dry herbs
- ½ teaspoon of paprika

Directions:

1. Preheat the Air Fryer to 380F.
2. Sprinkle the basket with oil.
3. Cut sausage in the pieces.
4. Rub them with salt, pepper, parsley, dry herbs and paprika.
5. Slice tomatoes in 4-6 pieces and blend with sausage.
6. Place in the Air Fryer and cook for 5 minutes at 380F.
7. Then beat 3 eggs, add to the combination.
8. Cook them for 5 minutes.
9. Serve hot with fresh vegetables or basil leaves.

Nutrition:

- Calories: 330
- Fat: 9,5g
- Carbohydrates: 8,7g
- Protein: 3,2g

Breakfast Potatoes

If you have some doubts about what prepare for breakfast – choose potatoes. You will like this meal because of crispy skin and also it is very tasty.

Prep time:10 minutes

Cooking time: 20-22 minutes

Servings:4

Ingredients:

- ¼ cup of onion
- 1 small red pepper
- ½ teaspoon of salt
- ½ teaspoon of pepper
- 2 tablespoon of oil
- 4 big potatoes
- ½ teaspoon of dry herbs

Directions:

1. Wash and chop potatoes in the pieces.
2. Then add salt, pepper and dry herbs to the bowl with potatoes and blend everything well.
3. Preheat the Air Fryer to 380F.
4. Sprinkle the frying basket with oil.
5. Place potatoes in the Air Fryer and cook for 15 minutes at 380F.
6. Then chop pepper in the strings.
7. Chop onion in the pieces and blend with pepper.
8. Add this combination to potatoes and cook for 5-7 minutes more at 320F.
9. Serve hot with sauces.

Nutrition:

- Calories: 190
- Fat: 6,7g
- Carbohydrates: 5,3g
- Protein: 2,3g

Fried English Breakfast

Do you want to prepare real English breakfast at home? There is nothing easier then to prepare it. You will like it, be sure.

Prep time:10 minutes

Cooking time: 16 minutes

Servings:4

Ingredients:

- 8 pieces of toast
- 1 can backed beans
- 4 eggs
- 8 pieces of bacon
- 8 medium sausages
- 2 tablespoons of oil

Directions:

1. Preheat the Air Fryer to 320F.
2. Sprinkle the frying basket with oil.
3. Then put sausages in the Air Fryer and cook for 5 minutes.
4. After that cook bacon for 5 minutes at the same temperature.
5. Then cook some part of sausages and bacon for 5 minutes at 300F.
6. The second part of bacon and sausages should be cooked with egg for 6 minutes at 300F.
7. Serve hot with sauce and vegetables.
8. You can enjoy with this meal with your friends.

Nutrition:

- Calories: 287
- Fat: 4,9g
- Carbohydrates: 4,9g
- Protein: 1,7g

Eggs with Bacon and Buns

It is possible to prepare delicious buns for breakfast. They need only a few minutes of your time and your breakfast is ready.

Prep time:10 minutes

Cooking time: 7 minutes

Servings:4

Ingredients:

- 4 cooked eggs
- 8 pieces of bacon
- ½ teaspoon of salt
- ½ teaspoon of pepper
- 4 buns
- 2 tablespoons of butter
- 2 tablespoons of ketchup
- ½ teaspoon of dry herbs
- 2 tablespoons of mustard

Directions:

1. Preheat the Air Fryer to 320F.
2. Sprinkle the Air Fryer basket with oil.
3. Then cut buns in 2 parts.
4. Cover them with butter and cook in the Air Fryer at 320F for 2 minutes.
5. Then mix sauce, mustard, salt, pepper, dry herbs in the bowl.
6. Cover buns with this mixture.
7. Put the piece of bacon and egg on buns.
8. Then cook them in the Air Fryer at 320F for 5 minutes.
9. Serve hot and enjoy with your breakfast.

Nutrition:

- Calories: 213
- Fat: 5,7g
- Carbohydrates: 4,6g
- Protein: 1,2g

Scrumptious and Sharp Burrito

Try to prepare this dish for your family. Very appetizing and satisfying! Remember this recipe!

Prep time:10 minutes

Cooking time: 25 minutes

Servings:4

Ingredients:

- 1 tortilla
- ½ teaspoon of salt
- ½ teaspoon of pepper
- ¼ cup of cheese
- ¼ avocado
- 1/3 big red pepper
- 2 eggs
- 3-4 piece of chicken fillets
- 2 tablespoons of oil

Directions:

1. Beat 2 eggs in the bowl.
2. Then add salt and pepper and mix everything.
3. Sprinkle the Air Fryer basket with oil.
4. Place eggs in the Air Fryer and cook for 5 minutes at 290F.
5. Slice fillets in the pieces.
6. Then cut avocado and red pepper and blend together.
7. Blend all ingredients and add to eggs.
8. Cook for 15 minutes at 330F.
9. Then cover with cheese and cook for 5 minutes more at 300F.
10. Serve hot and enjoy with your meal.

Nutrition:

- Calories: 187
- Fat: 5,4g
- Carbohydrates: 4,2g
- Protein: 3,8g

Scrumptious Sausage with Cheese

Delicious and appetizing dish in a short time. When you prepare it, you will see that it is ideal meal for breakfast.

Prep time:10 minutes

Cooking time: 16 minutes

Servings:4

Ingredients:

- 5 tablespoons of ketchup or sauces
- 8 wooden skewers
- 1 can of roll dough
- 1 cup of cheese
- 8 sausages
- ½ teaspoon of salt
- ½ teaspoon of pepper

Directions:

1. Preheat the Air Fryer to 340F.
2. Sprinkle the basket with oil.
3. Then cut sausages in the pieces and mix them with salt and pepper.
4. Place them in the Air Fryer and cook for 3 minutes.
5. Then put on another side and cook for 3 minutes more at the same temperature.
6. Cut the dough in the pieces.
7. Blend sausages with ketchup.
8. Then put them on the pieces of dough.
9. Cover with cheese and make rolls.
10. Put in the Air Fryer and cook for 5 minutes at 300F.
11. Then place on the other side and cook for 5 minutes more.
12. The temperature should be the same.
13. Serve hot and enjoy with breakfast.

Nutrition:

- Calories: 210
- Fat: 5,9g
- Carbohydrates: 4,6g
- Protein: 3,9g

French Toasts Sticks

Such a bright, interesting toast is an excellent choice for breakfast. Also, it is wonderful and easy in preparing.

Prep time:10 minutes

Cooking time: 8 minutes

Servings:2

Ingredients:

- 2 teaspoons of icing sugar
- 2 teaspoons of maple syrup
- ½ teaspoon of cinnamon
- ½ teaspoon of nutmeg
- ½ teaspoon of salt
- 2 eggs
- 2 tablespoons of butter
- 4 pieces of bread
- 2 tablespoons of oil

Directions:

1. Sprinkle the frying basket with oil.
2. Then preheat the Air Fryer for 5 minutes to 340F.
3. Beat 2 eggs in the bowl.
4. Put salt, cinnamon and nutmeg there and blend everything well.
5. Cut bread in the strings.
6. Cover them with butter on 2 sides.
7. Place every bread string in egg.
8. Check if they are completely covered with egg.
9. Then place the strings in the Air Fryer and cook for 4 minutes.
10. Shake them well and cook for 4 minutes more at the same temperature.
11. Serve hot with the icing sugar or maple syrup.

Nutrition:

- Calories: 156
- Fat: 2,3g
- Carbohydrates: 3,2g
- Protein: 2,2g

Souffle for Breakfast

You can think that it is ordinary recipe, but when you try it, you will change your mind. You will remember this breakfast forever, because it is really delicious.

Prep time:15 minutes

Cooking time: 8 minutes

Servings:2

Ingredients:

- ½ teaspoon of parsley
- 2 tablespoons of cream
- 1 teaspoon of red pepper
- 2 eggs
- 2 tablespoons of oil

Directions:

1. Sprinkle the Air Fryer basket with oil.
2. Preheat it for 5 minutes to 300F.
3. Chop parsley and pepper in small pieces.
4. Beat eggs and mix them with parsley, cream and pepper.
5. Then put the mixture in the special forms for baking.
6. Cook for 8 minutes at 300F.
7. When they are ready, put them on the plate and leave for 10 minutes.
8. Serve warm with juice or tea.
9. Enjoy with this perfect breakfast.

Nutrition:

- Calories: 123
- Fat: 2,1g
- Carbohydrates: 3,2g
- Protein: 2,1g

Lunch

Salad with Cucumber and Potato

Incredibly delicious salad, which can be great for lunch. The combination of cucumber and potato is very tasty.

Prep time: 10 minutes

Cooking time: 25 minutes

Servings: 4

Ingredients:

- 1 tablespoon of oil
- 2 tablespoons of white wine vinegar
- 1 lb of potatoes
- 1 big cucumber
- 1/3 teaspoon of salt
- 1 big onion
- ½ teaspoon of pepper
- ½ teaspoon of cumin powder

Directions:

1. Wash and clean potatoes.
2. Then chop in the small pieces.
3. Sprinkle the Air Fryer with oil.
4. Preheat it to 350F for 2 minutes.
5. Place potatoes in the Air Fryer and cook for 20 minutes.
6. After that mix potatoes and cook for 5 minutes more.
7. Cut cucumber in the pieces.
8. Chop onion in the rings.
9. Add salt, pepper, cumin powder and vinegar.
10. Mix salad well.
11. Serve hot potatoes with salad.

Nutrition:

- Calories: 127
- Fat: 2,2g
- Carbohydrates: 6,9g
- Protein: 2,1g

Pepper Rice

This rice with spices is exactly for lunch. It is uncomplicated and flavorsome. Cook and see.

Prep time: 20 minutes

Cooking time: 25 minutes

Servings: 4

Ingredients:

- 2 cups of rice
- 3 cups of water
- ½ teaspoon of salt
- ½ teaspoon of pepper
- ½ teaspoon of garlic powder
- 1/3 teaspoon of onion powder
- 1/3 teaspoon of ginger
- 1/5 teaspoon of oregano
- 1/3 teaspoon of red chili pepper
- 2 tablespoons of oil

Directions:

1. Sprinkle the Air Fryer with oil.
2. Preheat it to 350F.
3. Put rice and water in the Air Fryer.
4. Add pepper, garlic powder, salt, onion powder, oregano and red chili pepper and blend everything well with rice.
5. Cook rice for 20 minutes.
6. After that blend it well and cook for 5 minutes.
7. Leave it for 20 minutes.
8. Serve hot with vegetables or meat.
9. Enjoy with peppery food.

Nutrition:

- Calories: 49
- Fat: 1,1g
- Carbohydrates: 3,9g
- Protein: 2,1g

Rice with Pepper

It will be tasty lunch. The red and green peppers will add special aroma to rice.

Prep time: 10 minutes

Cooking time: 25 minutes

Servings: 4

Ingredients:

- 1 lb of rice
- ½ teaspoon of pepper
- ½ teaspoon of salt
- 1/3 teaspoon of oregano
- 1/3 teaspoon of red chili pepper
- 1 big sweat red pepper
- 1 big green pepper
- 2 tablespoons of oil

Directions:

1. Wash rice under cold water.
2. Then grease the Air Fryer with oil.
3. Preheat it to 300F for 5 minutes.
4. After that place rice with water in the Air Fryer.
5. Cook it for 10 minutes.
6. Then add oregano, red chili pepper, salt and pepper and blend everything well.
7. Cook for 5 minutes.
8. Chop green and red peppers in the pieces.
9. Add them to rice and mix all components.
10. Cook the meal for 10 minutes.
11. Serve hot with sauces.

Nutrition:

- Calories: 89
- Fat: 1,3g
- Carbohydrates: 7,5g
- Protein: 2,4g

Piquant Fried Pumpkin

It is forceful and appetizing meal. Just prepare and try. This meal is really flavorsome.

Prep time: 15 minutes

Cooking time: 20 minutes

Servings: 4

Ingredients:

- 2 tablespoons of olive oil
- ½ teaspoon of pepper
- ½ teaspoon of salt
- 2 teaspoons of oregano
- 2 teaspoons of garlic powder
- ½ lb of pumpkin

Directions:

1. Sprinkle the Air Fryer with 1 tablespoon of oil.
2. Preheat it to 350F for 5 minutes.
3. Wash and clean pumpkin.
4. Then chop it in the slices.
5. Take the container and blend pepper, salt, oregano, garlic powder with oil there.
6. Then rub pumpkin with seasonings.
7. Cook for 10 minutes.
8. Then blend everything well. Cook for 10 minutes again.
9. Serve hot with parsley or with different sauces.
10. Enjoy with appetizing meal.

Nutrition:

- Calories: 76
- Fat: 3,6g
- Carbohydrates: 5,2g
- Protein: 1,4g

Fried Zucchini with Almonds

It is appetizing meal, which you will like. Uncomplicated and fast – just cook and check it.

Prep time: 10 minutes

Cooking time: 5 minutes

Servings: 4

Ingredients:

- ½ teaspoon of salt
- ½ teaspoon of pepper
- ½ lb of zucchini
- 2 tablespoons of silvered almonds
- 2 tablespoons of oil
- 1/3 teaspoon of garlic powder
- ¼ teaspoon of oregano

Directions:

1. Sprinkle the Air Fryer with 1 tablespoon of oil.
2. Preheat it to 280F for 5 minutes.
3. Place almonds with the rest of oil in the bowl.
4. Chop zucchini in the pieces.
5. Add salt, pepper, garlic powder and oregano in the bowl.
6. Blend all components.
7. Place the mixture in the Air Fryer.
8. Cook this meal for 5 minutes.
9. Serve hot meal decorated with fresh basil leaves.
10. Eat appetizing meal!

Nutrition:

- Calories: 98
- Fat: 3,2g
- Carbohydrates: 5,4g
- Protein: 1,3g

Flavorsome Spaghetti with Backed Corn

It is incredible mixture, which everyone will like. Easy and simple.

Prep time: 10 minutes

Cooking time: 25 minutes

Servings: 4

Ingredients:

- ¼ cup of margarine
- 1 In of cream-style corn
- 1 cup of cheese
- 1 lb of whole kernel corn
- 1 cup of spaghetti
- ½ teaspoon of salt
- ½ teaspoon of pepper
- ½ teaspoon of oregano
- 2 teaspoons of oil

Directions:

1. Sprinkle the Air Fryer with oil.
2. Preheat it to 390F for 5 minutes.
3. Take the deep bowl.
4. Place spaghetti, cream-style corn and whole kernel corn there.
5. Blend everything well.
6. Then place the combination in the Air Fryer and cook for 15 minutes.
7. Grate cheese and add to the meal.
8. Cook for 10 minutes more.
9. Serve hot with sauces.
10. Enjoy with the appetizing and peppery meal.

Nutrition:

- Calories: 189
- Fat: 4,1g
- Carbohydrates: 8,9g
- Protein: 1,7g

Backed Cabbage

This lunch is really healthy. Also, cabbage is flavorsome and wonderful.

Prep time: 5 minutes

Cooking time: 40 minutes

Servings: 4

Ingredients:

- ½ teaspoon of salt
- ½ teaspoon of pepper
- 1 big cabbage
- 3 tablespoons of balsamic vinegar
- 2 tablespoons of oil
- 1 green or red cabbage
- 1/3 teaspoon of cumin powder
- 2 tablespoons of olive oil

Directions:

1. Grease the Air Fryer with oil.
2. Then preheat it to 350F for 2 minutes.
3. Cut cabbage in the pieces.
4. Then grease with olive oil and add salt, pepper and cumin powder.
5. Blend the components well.
6. Put cabbage in the Air Fryer.
7. Cover with the foil.
8. Cook for 20 minutes.
9. Then add balsamic vinegar, blend cabbage well and cover with the foil again.
10. Cook for 20 minutes more.
11. Serve hot with meat and parsley or dill.
12. Enjoy with the appetizing meal.

Nutrition:

- Calories: 67
- Fat: 1,8g
- Carbohydrates: 6,5g
- Protein: 1,2g

Scrumptious Sour Pickled Mustard Greens

It is surely uncomplicated and yummy lunch. This method is unusual and does not need a lot of time to be cooked.

Prep time: 5 minutes

Cooking time: 6-10 minutes

Servings: 4

Ingredients:

- ½ teaspoon of salt
- ½ teaspoon of pepper
- 1/3 cup of water
- 2 tablespoons of oil
- 6 oz of bean sprouts
- 8 oz of sour pickled mustard greens
- 4 small onions
- 2 teaspoons of fish sauce

Directions:

1. Sprinkle the Air Fryer with oil.
2. Preheat it for 3 minutes to 360F.
3. Put bean sprouts in the Air Fryer and cook for 1-2 minutes.
4. Then chop onions and cook them with mustard in the Air Fryer for 3-4 minutes.
5. After that add salt and pepper.
6. Then mix the ingredients, add water and cook for 2-3 minutes more.
7. Serve hot with basil leaves and lemon.
8. Enjoy with the tempting and vigorous meal.

Nutrition:

- Calories: 112
- Fat: 2,1g
- Carbohydrates: 7,2g
- Protein: 1,4g

Fried Kimchi and Bean Sprouts

It is healthy and peppery food. You will like it because it has incredibly tasty ingredients.

Prep time: 10 minutes

Cooking time: 5 minutes

Servings: 2

Ingredients:

- 1 teaspoon of oil
- 7 oz of bean sprouts
- 4-5 pieces of the kimchi
- 4 tablespoons of oil
- 4 onions
- ½ teaspoon of salt
- ½ teaspoon of oil
- 1 teaspoon of sesame oil

Directions:

1. Sprinkle the Air Fryer with oil.
2. Preheat it to 350F for 2-3 minutes.
3. Cook bean sprouts for 1-2 minutes in the Air Fryer.
4. Chop onions and kimchi in the pieces.
5. After that add them to bean sprouts.
6. Then blend all ingredients and add salt, sesame oil and pepper.
7. Add water and cook the products in the Air Fryer for 3-4 minutes.
8. Serve hot with meat or sauces.
9. Enjoy with tasty meal with your friends.

Nutrition:

- Calories: 99
- Fat: 2,2g
- Carbohydrates: 7g
- Protein: 1,3g

Spring Onion for Barbecue

Even if this is for the picnic, you can prepare it for lunch at home. It is appetizing, fast and delicious.

Prep time: 40 minutes

Cooking time: 10 minutes

Servings: 4

Ingredients:

- 1-2 lemons
- ½ teaspoon of salt
- ½ teaspoon of pepper
- 3 tablespoons of olive oil
- 1 bunch of spring onion

Directions:

1. Wash spring onion.
2. Then cut it in the pieces.
3. Put in the bowl.
4. After that rub it with oil, salt and pepper.
5. Leave for 30 minutes.
6. Sprinkle the Air Fryer with oil.
7. Preheat it for 3 minutes to 350F.
8. Then sprinkle onion with juice of lemon.
9. Put in the Air Fryer and cook for 10 minutes.
10. Serve hot with potatoes or meat.
11. Decorate with basil leaves or mint.
12. It is possible to serve with sauce or ketchup.
13. Eat appetizing meal and enjoy.

Nutrition:

- Calories: 46
- Fat: 1,1g
- Carbohydrates: 2,7g
- Protein: 0,5g

Fried Rice with Pasta

It is the great and healthy meal. Enjoy with this tasty rice.

Prep time: 5 minutes

Cooking time: 30 minutes

Servings: 4

Ingredients:

- ½ teaspoon of salt
- ½ teaspoon of pepper
- 4 chicken stock cubes
- 2 cups of chicken stock
- 3 cups of rice
- 1 cup of spaghetti
- 3 oz of butter
- 2 teaspoons of oil
- ½ teaspoon of oregano

Directions:

1. Sprinkle the Air Fryer with oil.
2. Then preheat it to 350F for 3 minutes.
3. Put butter with spaghetti in the Air Fryer.
4. Add rice and chicken stock.
5. Then add chicken stock cubes and blend everything well.
6. Add salt, pepper and oregano.
7. Cook for 15 minutes.
8. Then blend everything well and cook for 15 minutes more.
9. Serve hot with sauces.
10. Decorate with vegetables or basil leaves.
11. Enjoy with the tasty meal.

Nutrition:

- Calories: 148
- Fat: 1,9g
- Carbohydrates: 3,1g
- Protein: 1g

Tasty and Pepper Pea Sprouts

This meal is fantastic. You will feel it from the first piece of the food. Cook and enjoy!

Prep time: 5 minutes

Cooking time: 10 minutes

Servings: 4

Ingredients:

- 1 teaspoon of sesame oil
- ½ teaspoon of salt
- ½ teaspoon of oil
- 1 teaspoon of sauce
- 5 cloves of garlic
- 2 teaspoons of oil
- 1 bag of pea sprouts

Directions:

1. Wash and clean pea sprouts.
2. Then dry them and leave aside.
3. Sprinkle the Air Fryer with oil
4. Then preheat it to 350F.
5. After that put pea sprouts and chopped garlic.
6. Add salt, pepper and mix everything.
7. Then add sesame oil and sauce.
8. Mix all ingredients completely.
9. Put everything in the Air Fryer and cook for 10 minutes.
10. Serve hot with vegetables or with different kinds of sauces.
11. Enjoy with this meal.

Nutrition:

- Calories: 98
- Fat: 2,1g
- Carbohydrates: 3,5g
- Protein: 1,1g

Puree with Pumpkin

It can be as your lunch as your breakfast. Easy, healthy and very delicious.

Prep time: 10 minutes

Cooking time: 1 hour and 10 minutes

Servings: 4

Ingredients:

- ½ teaspoon of pepper
- ½ teaspoon of salt
- ½ teaspoon of cumin powder
- 1 tablespoon of oil
- 2 small pumpkins
- 6-8 bay leaves
- ½ teaspoon of coriander
- 1-2 tablespoons of maple syrup

Directions:

1. Grease the Air Fryer with oil.
2. Preheat the Air Fryer to 300F for 2 minutes.
3. Wash and clean pumpkins.
4. After that set pumpkin in the Air Fryer.
5. Cook this meal for 1 hour.
6. Then add butter and bay leaves to pumpkin and cook for 10 minutes more.
7. Add salt and black pepper.
8. Pour pumpkin with maple syrup.
9. Serve hot with the orange or apple juice.
10. Enjoy with your perfect and appetizing lunch.

Nutrition:

- Calories: 89
- Fat: 1,3g
- Carbohydrates: 6,8g
- Protein: 1,1g

Tasty Fried Celeriac

Just prepare this meal one time. Tasty, delicious and not complicated for cooking. This meal is simply and fantastic.

Prep time: 15 minutes

Cooking time: 25 minutes

Servings: 4

Ingredients:

- ½ teaspoon of pepper
- ½ teaspoon of salt
- 1 teaspoon of parsley
- 1 bunch of spring onion
- 2 celeriac roots
- 2 tablespoon of oil
- 1 teaspoon of onion powder
- ½ teaspoon of garlic powder

Directions:

1. Sprinkle the Air Fryer with oil.
2. Preheat it to 350F.
3. Chop celeriac roots in the pieces.
4. Cook in the Air Fryer for 10 minutes.
5. Then cut spring onion in the pieces and add to celeriac roots.
6. Add salt and pepper and blend everything very well.
7. Cook for 15 minutes more.
8. After that decorate the meal with parsley or basil leaves.
9. Serve hot with the fresh vegetables.

Nutrition:

- Calories: 99
- Fat: 1,2g
- Carbohydrates: 6,6g
- Protein: 1g

Courgette With Lemon and Butter

It is interesting and appetizing meal. The ingredients are very simple and it is easy for preparing.

Prep time: 5 minutes

Cooking time: 15 minutes

Servings: 4

Ingredients:

- ½ teaspoon of salt
- ½ teaspoon of pepper
- ½ lemon
- 2 oz of butter
- 2 tablespoons of oil
- 4-6 courgettes
- ½ teaspoon of onion powder

Directions:

1. Rinse and clean courgettes.
2. Chop them in the food processor.
3. Then sprinkle the Air Fryer with oil.
4. Preheat it to 300F.
5. Then place courgettes in the Air Fryer.
6. Cook for 5 minutes.
7. Then add onion powder, salt, pepper and then blend the products well.
8. Sprinkle the meal with juice of lemon and cook for 10 minutes more.
9. Serve it hot with fresh basil leaves or with parsley.
10. Cook and enjoy!
11. This meal is really scrumptious.

Nutrition:

- Calories: 138
- Fat: 1,2g
- Carbohydrates: 5,3g
- Protein: 1,3g

Side Dishes

Peppery Bacon with Appetizing Brussels Sprouts and Sweat Cream

This method in uncommon and wonderful. Make this flavorsome meal.

Prep time: 10 minutes

Cooking time: 50 minutes

Servings:4

Ingredients:

- 1 tablespoon of thyme leaves
- ½ teaspoon of pepper
- ½ teaspoon of salt
- 2 teaspoons of oil
- 1 cup of milk
- 2 cups of cream
- 4 tablespoons of butter
- 4 tablespoons of flour
- 4 shallots
- 1 lb of bacon
- 1 lb of brussels sprouts

Directions:

1. Sprinkle the Air Fryer with oil.
2. Preheat to 400F.
3. Then place brussels sprouts in the Air Fryer and cook for 20 minutes.
4. Chop meat in the pieces and add to brussels sprouts.
5. Cook them for 20 minutes more.
6. Blend cream, pepper, salt, milk, butter, flour, shallots in the bowl.
7. Cover bacon with brussels sprouts with this combination.
8. Cook for 10 minutes more.
9. Serve hot with basil leaves or mint.
10. Enjoy with the meal.

Nutrition:

- Calories: 137
- Fat: 2,2g
- Carbohydrates: 7,3g
- Protein: 1,1g

Browned Pasta Salad with Vegetables

It is really scrumptious and uncomplicated salad. Just prepare and you will like it forever.

Prep time: 10 minutes

Cooking time: 25 minutes

Servings:4

Ingredients:

- ¼ cup of oil
- 1 teaspoon of fresh basil
- 3 tablespoons of vinegar
- 1 teaspoon of dry herbs
- 1 cup of tomatoes
- 1 onion
- 4 oz of mushrooms
- 1 zucchini
- 1 yellow squash
- 1 red pepper
- 1 green pepper
- 1 yellow pepper
- ½ teaspoon of red chili pepper

Directions:

1. Chop yellow, red and green peppers in the pieces.
2. Then slice onion and add there.
3. After that cut tomatoes in the pieces.
4. Blend the ingredients.
5. After that chop zucchini and mushrooms.
6. Then blend everything well.
7. Add red chili pepper, salt, chopped squash, dry herbs, vinegar and basil.
8. Blend salad.
9. Sprinkle the Air Fryer with oil.
10. Preheat to 350F.
11. Cook salad for 15 minutes.
12. Then blend everything well and cook for 10 minutes more.
13. Serve hot.

Nutrition:

- Calories: 79
- Fat: 1,1g
- Carbohydrates: 8,4g
- Protein: 1g

Potato Chips with Cream and Onion

It is easy and wonderful meal. You can make it any time you wish, because it does not require a lot of your efforts.

Prep time: 10 minutes

Cooking time: 15 minutes

Servings:4

Ingredients:

- 2 big potatoes
- 3 tablespoons of oil
- ½ teaspoon of salt
- ½ teaspoon of pepper
- ½ cup of sour cream
- 1 teaspoon of lemon juice
- 1/3 teaspoon of red chili pepper
- 1/3 teaspoon of paprika
- ½ onion

Directions:

1. Rinse and clean potatoes.
2. Then chop potatoes in the small pieces.
3. Rub potatoes with salt, pepper, red chili pepper and paprika.
4. Sprinkle the Air Fryer with oil.
5. Preheat it to 350F.
6. Cook chips for 5 minute.
7. Then shake them well and cook for 10 minutes more.
8. Then blend sour cream with lemon juice.
9. Chop onion in small pieces and add to sour cream.
10. Blend everything well.
11. Place chips in sour cream and eat.

Nutrition:

- Calories: 123
- Fat: 1,6g
- Carbohydrates: 4,3g
- Protein: 1,1g

The Biscuits with Cheese

These biscuits are appetizing. Do not miss your chance to cook them.

Prep time: 10 minutes

Cooking time: 20 minutes

Servings:4

Ingredients:

- 1 cup of flour
- 1 tablespoon of butter
- 3 tablespoons of oil
- 1 cup of buttermilk
- ½ cup of butter
- ½ cup of cheese
- 2 cups of self-rising flour
- 2 tablespoons of sugar

Directions:

1. Blend flour with butter.
2. Then add buttermilk and blend everything.
3. Add sugar, self-rising flour and butter to the ingredients.
4. Blend the components well.
5. Then chop cheese in the small pieces and add to the mixture.
6. Sprinkle the Air Fryer with oil.
7. Preheat it to 350F.
8. Make the balls and cook them in the Air Fryer for 10 minutes.
9. Then shake them well and cook for 10 minutes more.
10. Serve hot with vegetables.

Nutrition:

- Calories: 114
- Fat: 2,3g
- Carbohydrates: 4,6g
- Protein: 1,2g

Browned Zucchini

This meal is flavorsome and wonderful. Just cook and try it.

Prep time: 10 minutes

Cooking time: 20 minutes

Servings:4

Ingredients:

- 1 big zucchini
- ½ teaspoon of salt
- ½ teaspoon of pepper
- 2 eggs
- ½ cup of bread crumbs
- ½ cup of flour
- ½ cup of mayonnaise
- 1 teaspoon of lemon juice
- 1 teaspoon of garlic powder

Directions:

1. Rinse and chop zucchini in the pieces.
2. Beat eggs in the bowl.
3. Then add salt, pepper, lemon juice and garlic powder to eggs and blend them.
4. Take zucchini and place in flour.
5. Then place in eggs and in mayonnaise.
6. After that cover with bread crumbs
7. Sprinkle the Air Fryer with oil.
8. Preheat it to 300F for 3 minutes.
9. Cook zucchini for 10 minutes.
10. Then shake them. Cook for 10 minutes more.
11. Serve hot with pasta.
12. Decorate with parsley.

Nutrition:

- Calories: 103
- Fat: 2,2g
- Carbohydrates: 4,2g
- Protein: 1,1g

Appetizing Tomatoes with Garlic

If you like something peppery, choose this recipe. You will be delighted with the result.

Prep time: 10 minutes

Cooking time: 15 minutes

Servings:4

Ingredients:

- ½ teaspoon of pepper
- ½ teaspoon of salt
- 2 teaspoons of oil
- 4 tomatoes
- ½ teaspoon of dry herbs
- ½ teaspoon of dried thyme
- 1 onion
- ½ teaspoon of oregano
- ½ teaspoon of red chili pepper
- 1/3 teaspoon of parsley

Directions:

1. Sprinkle the Air Fryer with oil.
2. Preheat to 350F.
3. Wash and chop tomatoes.
4. Then slice onion in the rings.
5. Rub tomatoes with salt, pepper, dry herbs, thyme, oregano, red chili pepper and parsley.
6. Place the rings of onion on tomatoes.
7. Cook them for 15 minutes in the Air Fryer.
8. Serve hot with parsley or basil leaves.
9. Eat this meal and you will like it very much.

Nutrition:

- Calories: 89
- Fat: 1,1g
- Carbohydrates: 4,2g
- Protein: 1g

Appetizing Potatoes

If you like crispy and flavorsome potatoes – then this technique is for you. Cook and enjoy with the fiery meal.

Prep time: 10 minutes

Cooking time: 25 minutes

Servings:4

Ingredients:

- 4 potatoes
- ½ teaspoon of salt
- ½ teaspoon of pepper
- 2 tablespoons of oil
- ½ teaspoon of rosemary
- 2 basil leaves
- ½ teaspoon of dry herbs
- 1/3 teaspoon of cumin
- ½ teaspoon of red chili pepper

Directions:

1. Rinse and clean potatoes
2. Cut potatoes in the strings.
3. Then rub them with pepper, rosemary, dry herbs, cumin and red chili pepper.
4. Sprinkle the Air Fryer with oil.
5. Preheat it to 360F.
6. Place potatoes in the Air Fryer and cook for 10 minutes.
7. Then shake well and cook for 15 minutes more.
8. Serve hot with sauces and vegetables.
9. Eat and enjoy with your family.

Nutrition:

- Calories: 114
- Fat: 1,2g
- Carbohydrates: 2,4g
- Protein: 1g

Fried Scrumptious Snow Pea

It is easy and lovely meal. Cook and enjoy.

Prep time: 5 minutes

Cooking time: 23 minutes

Servings:4

Ingredients:

- 1 lb of snow peas
- ½ teaspoon of pepper
- ½ teaspoon of salt
- 2 tablespoons of oil
- ½ teaspoon of dry herbs
- ½ teaspoon of rosemary
- 3 cloves of garlic
- ½ teaspoon of paprika
- 1/3 oregano
- 1 teaspoon of dill
- 1 teaspoon of parsley

Directions:

1. Wash and clean snow peas.
2. Then rub them with pepper, salt, dry herbs, rosemary, dill, parsley and paprika.
3. Chop garlic in the pieces and add there.
4. Blend everything well.
5. Sprinkle the Air Fryer with oil.
6. Preheat to 300F.
7. Cook the meal for 10 minutes.
8. Then blend everything again and cook for 13 minutes more.
9. Serve hot meal with fresh parsley.
10. Eat the appetizing food.

Nutrition:

- Calories: 98
- Fat: 1,3g
- Carbohydrates: 2,7g
- Protein: 1,1g

Tempting Smished Potatoes

Uncomplicated, easy and flavorsome. You should prepare this meal because of the yummy aroma.

Prep time: 5 minutes

Cooking time: 25 minutes

Servings:4

Ingredients:

- 1 teaspoon of pepper
- ½ teaspoon of salt
- 6 potatoes
- 1 green pepper
- ½ teaspoon of dry herbs
- ½ teaspoon of cumin
- 1/3 teaspoon of paprika
- 1/3 teaspoon of red chili pepper
- 3 cloves of garlic

Directions.

1. Wash and clean potatoes.
2. Slice potatoes in the pieces.
3. Blend pepper, salt, dry herbs, cumin, paprika, red chili pepper and garlic in the bowl.
4. Rub potatoes with these spices.
5. Then chop pepper in the strings.
6. Sprinkle the Air Fryer with oil.
7. Preheat it to 300F.
8. Cook potatoes for 15 minutes.
9. Then shake well, add pepper and cook for 10 minutes more.
10. Serve hot with sauce, basil and parsley.

Nutrition:

- Calories: 114
- Fat: 1,2g
- Carbohydrates: 2,8g
- Protein: 1g

Flavorsome Potatoes Pesto

If you are looking for something original and simple, this meal is exactly for you. It will not take too much of your time for preparing.

Prep time: 10 minutes

Cooking time: 25 minutes

Servings:4

Ingredients:

- 4 potatoes
- ½ teaspoon of pepper
- ½ teaspoon of salt
- 2 teaspoons of oil
- 1 tablespoon of pesto
- ½ teaspoon of garlic powder
- ½ teaspoon of onion powder
- ¼ cup of milk

Directions:

1. Wash and clean potatoes.
2. Cut potatoes in the pieces.
3. Rub them with salt, pepper, garlic powder and onion powder.
4. Sprinkle the Air Fryer with oil.
5. Then preheat the Air Fryer to 350F.
6. Blend potatoes with milk and pesto.
7. Cook in the Air Fryer for 10 minutes.
8. Then blend everything well and cook for 15 minutes more.
9. Serve hot with basil leaves.

Nutrition:

- Calories: 106
- Fat: 1,2g
- Carbohydrates: 2,7g
- Protein: 1g

Appetizing Salad with Warm Cabbage and Pasta

It is easy and pleasant meal. You can prepare it as for the holiday as for supper.

Prep time: 15 minutes

Cooking time: 25 minutes

Servings:4

Ingredients:

- 4 oz of cooked pasta
- ½ teaspoon of salt
- ½ teaspoon of pepper
- 2 tablespoons of oil
- ½ teaspoon of dry herbs
- ½ teaspoon of paprika
- 1/3 teaspoon of red chili pepper
- 1/3 teaspoon of cumin
- 3 cloves of garlic
- ½ teaspoon of onion powder
- ½ cabbage
- 1 onion

Directions:

1. Boil pasta for 10 minutes.
2. Then place pasta in the bowl.
3. Add onion powder, chopped garlic, paprika, red chili pepper, salt, pepper, dry herbs and blend everything.
4. Then chop onion in the rings.
5. Add them to pasta.
6. Cut cabbage in the strings and blend everything.
7. Sprinkle the Air Fryer with oil.
8. Preheat the Air Fryer to 350F.
9. Place pasta in the Air Fryer and cook for 15 minutes.
10. Then blend everything and cook for 5 minutes more.
11. Serve hot with chili sauce.

Nutrition:

- Calories: 156
- Fat: 1,3g
- Carbohydrates: 3,3g
- Protein: 1g

Backed Chips with Zucchini

It is fantastic and easy meal. Just cook and you will be delighted.

Prep time: 10 minutes

Cooking time: 20 minutes

Servings:4

Ingredients:

- ½ cup of bread crumbs
- ½ teaspoon of salt
- ½ teaspoon of pepper
- 2 tablespoons of oil
- 3 tablespoons of grated cheese
- ½ teaspoon of dry herbs
- 1/3 teaspoon of red chili pepper
- 1/3 teaspoon of paprika
- 2 eggs
- 2 big zucchinis

Directions:

1. Cut zucchinis in the pieces.
2. Then rub them with salt, pepper, dry herbs, red chili pepper and paprika.
3. Sprinkle the Air Fryer with oil.
4. Preheat it to 300F.
5. Then beat eggs.
6. Place zucchinis in eggs and after that in bread crumbs.
7. Cook them in the Air Fryer for 10 minutes.
8. Then shake them well and cook for 10 minutes more.
9. Serve hot with sauces and basil leaves.

Nutrition:

- Calories: 121
- Fat: 1,2g
- Carbohydrates: 3,2g
- Protein: 1g

Appetizing Fried Sweat Carrot

If you need to make something easy and pleasant, cook this carrot. The meal is scrumptious and fantastic.

Prep time: 10 minutes

Cooking time: 20 minutes

Servings:4

Ingredients:

- 1 lb of carrot
- ½ teaspoon of pepper
- ½ teaspoon of salt
- 2 tablespoons of oil
- ½ teaspoon of dry herbs
- ½ teaspoon of cumin
- 1/3 teaspoon of red chili pepper
- ½ teaspoon of rosemary
- ½ teaspoon of paprika
- 1 tablespoon of butter
- 1 teaspoon of parsley

Directions:

1. Sprinkle the Air Fryer with oil.
2. Then preheat it for 5 minutes to 320F.
3. Wash and clean carrot.
4. Then chop it in the strings.
5. Blend the herbs, salt, pepper, cumin, red chili pepper, paprika, rosemary and parsley in the bowl.
6. Rub carrot with these spices.
7. Then place carrot in the Air Fryer and cook for 10 minutes.
8. Shake and cook the food for 10 minutes more.
9. Serve hot with sauces.

Nutrition:

- Calories: 89
- Fat: 1,3g
- Carbohydrates: 3,5g
- Protein: 1g

Peppery Rice with Herbs

This rice will be great with meat. It is very flavorsome, piquant and delicious.

Prep time: 10 minutes

Cooking time: 30 minutes

Servings:4

Ingredients:

- 1 cup of rice
- ½ teaspoon of pepper
- ½ teaspoon of salt
- 2 tablespoons of oil
- 2 cups of water
- 1 onion
- ½ teaspoon of red chili pepper
- ½ teaspoon of paprika
- ½ teaspoon of dry herbs
- ¼ tablespoon of parsley

Directions:

1. Sprinkle the Air Fryer with oil.
2. Preheat it to 350F.
3. Blend rice with pepper, salt, paprika, red chili pepper, dry herbs and parsley.
4. Then add water.
5. Cook it in the Air Fryer for 15 minutes.
6. Then chop onion in the pieces.
7. Add to rice and blend everything well.
8. Cook for 15 minutes more.
9. Serve hot with vegetables.
10. Enjoy with flavorsome and scrumptious meal.

Nutrition:

- Calories: 68
- Fat: 1,1g
- Carbohydrates: 2,3g
- Protein: 1g

Flavorsome Tomatoes with Cheese

It is very easy for the preparing. Also, it is scrumptious. Cook and you will get the great result.

Prep time: 10 minutes

Cooking time: 10 minutes

Servings:4

Ingredients:

- 10 tomatoes
- ½ teaspoon of pepper
- ½ teaspoon of salt
- 3 tablespoons of oil
- 1/3 teaspoon of red chili pepper
- ½ tablespoon of dill
- 1/3 tablespoon of parsley
- ½ teaspoon of cumin
- ½ teaspoon of oregano
- 3 cloves of garlic
- 3 tablespoons of grated cheese

Directions:

1. Wash and chop tomatoes in the pieces.
2. Then blend pepper, salt, dill, red chili pepper, parsley, oregano, cumin and chopped garlic in the bowl.
3. Rub tomatoes with this mixture.
4. Sprinkle the Air Fryer with oil.
5. Preheat it to 300F.
6. Cook tomatoes for 5 minutes.
7. Then cover them with cheese and cook for 5 minutes more.
8. Serve hot with sauce.

Nutrition:

- Calories: 71
- Fat: 1,1g
- Carbohydrates: 3,5g
- Protein: 1g

Main dishes

Chicken Curry

You even cannot imagine how flavorsome it is. Your guests will be surprised when they try this meal.

Prep time: 15 minutes

Cooking time: 20 minutes

Servings: 4

Ingredients:

- ½ teaspoon of salt
- ½ teaspoon of pepper
- 2 tablespoons of oil
- 5 tablespoons of garlic
- 2 tomatoes
- 3 tablespoons of cream
- 4 cloves of garlic
- 1 ginger
- 1 green chili
- 1 tablespoon of garam masala
- 3 onions
- ½ teaspoon of turmeric
- ½ teaspoon of red chili pepper
- 2 black cardamom
- 4 green cardamom
- 2 bay leaves
- 2 lbs of chicken fillets

Directions:

1. Wash and cut chicken fillet in the pieces.
2. After that blend salt, pepper, garlic, ginger, garam masala, turmeric, red chili pepper, cardamom and bay leaves together in the bowl.
3. Rub the pieces of chicken with these spices.
4. Sprinkle the Air Fryer with oil.
5. Preheat it to 350F for 5 minutes.
6. Place chicken in the Air Fryer and cook for 5 minutes.
7. Chop tomatoes and green chili with onions.
8. Add them to the rest of the products.
9. Then add cream and blend.
10. Cook for 15 minutes more.
11. Serve hot with sauces.

Nutrition:

- Calories: 287
- Fat: 8,9g
- Carbohydrates: 10,3g
- Protein: 5,6g

Tasty and Spicy Chicken Stew

When you taste this meal for the first time, you will like it. Delicious and spicy, it is very tasty.

Prep time: 15 minutes

Cooking time: 30 minutes

Servings: 4

Ingredients:

- ½ teaspoon of salt
- ½ teaspoon of pepper
- 2 tablespoons of oil
- 1 tablespoon of flour
- 4-5 corns
- 3 oz of mushrooms
- 4 cloves of garlic
- 1 onion
- ½ red capsicum
- ½ green capsicum
- 1 broccoli
- 1 carrot
- 5-7 beans
- ¼ teaspoon of rosemary
- 2 tablespoons of oil
- 7 oz of chicken fillet

Directions:

1. Wash and clean chicken fillet.
2. Then chop it in the pieces.
3. Mix salt, pepper, flour, garlic, rosemary in the bowl.
4. Mix chicken with these spices.
5. Then chop mushrooms, onion, carrot, broccoli and mix with chicken.
6. After that, add the corns and beans.
7. Mix all components again.
8. Sprinkle the Air Fryer with oil.
9. Preheat it to 350F.
10. Cook the mixture in the Air Fryer for 15 minutes.
11. Then mix everything and cook for 15 minutes more.
12. Serve hot with vegetables.

Nutrition:

- Calories: 156
- Fat: 7,3g
- Carbohydrates: 9,8g
- Protein: 4,6g

Piquant Curry with Mutton

This curry meal is perfect. All components are appetizing.

Prep time: 20 minutes

Cooking time: 40 minutes

Servings: 4

Ingredients:

- 2 tablespoons of oil
- ½ teaspoon of salt
- ½ teaspoon of pepper
- 4 tablespoon of mustard oil
- 1 tablespoon of garam masala
- 1 teaspoon of red chili pepper
- 1 teaspoon of cumin
- 1 teaspoon of turmeric
- 6 oz of onions
- 1 tablespoon of ginger
- 1 lb of tomatoes
- 2 lbs of mutton

Directions:

1. Grease the Air Fryer with oil.
2. Preheat it for 5 minutes to 350F.
3. Rinse and clean meat.
4. Cut it in the pieces.
5. Then rub meat with salt, pepper, mustard oil, garam masala, red chili pepper, cumin and turmeric.
6. Cut tomatoes, onions and ginger in the pieces and blend everything.
7. Cook the food for 20 minutes.
8. Then blend everything and cook for 20 minutes.
9. Serve hot with parsley.
10. Enjoy with the food.

Nutrition:

- Calories: 245
- Fat: 10,3g
- Carbohydrates: 9,7g
- Protein: 7,9g

Sharp and Yummy Chicken in Appetizing Yogurt Sauce

It is flavorsome meal. This chicken is tempting.

Prep time: 20 minutes

Cooking time: 30 minutes

Servings: 4

Ingredients:

- ½ tablespoon of coriander
- 2 tablespoons of oil
- ½ teaspoon of pepper
- ½ teaspoon of salt
- ¼ teaspoon of turmeric
- 2 tablespoons of mustard oil
- ½ tablespoon of cumin
- 1 cup of yogurt
- 2 tomatoes
- 5 cloves of garlic
- 1 teaspoon of red chili pepper
- 3 onions
- 2 lbs of chicken fillets

Directions:

1. Sprinkle the Air Fryer with oil.
2. Preheat it to 350F for 3 minutes.
3. Rinse and clean chicken.
4. Chop meat in the pieces.
5. Blend the coriander, pepper, salt, turmeric, mustard oil, cumin, garlic and red chili pepper in the bowl.
6. Then rub chicken pieces with spices.
7. Cook the poultry in the Air Fryer for 10 minutes.
8. Then chop tomatoes, onions in the pieces.
9. Blend everything, cover with yogurt. Cook for 20 minutes.
10. Serve it hot.
11. Decorate the appetizing chicken with parsley and basil leaves.

Nutrition:

- Calories: 245
- Fat: 10,3g
- Carbohydrates: 9,7g
- Protein: 7,9g

Wonderful Jackfruit with Vegetables

It you like vegetables, this meal will be perfect for you. Cook and enjoy with it.

Prep time: 30 minutes

Cooking time: 45 minutes

Servings: 4

Ingredients:

- 1 teaspoon of coriander
- ½ teaspoon of salt
- 2 tablespoons of oil
- ½ teaspoon of pepper
- 1 cup of yogurt
- 5 tablespoons of butter
- 7 oz of onion
- 1 garlic
- ½ teaspoon of red chili pepper
- ½ teaspoon of coriander
- 2 potatoes
- 8 oz of jackfruit

Directions:

1. Cut potatoes and jackfruit in the pieces and mix them.
2. Then rub these products with salt, pepper, butter, red chili pepper, coriander.
3. Sprinkle the Air Fryer with oil.
4. Preheat the Air Fryer to 300F for 2-3 minutes.
5. Cook the products for 20 minutes.
6. Then chop onion and add to potatoes.
7. Cover everything with yogurt and cook for 25 minutes more.
8. Serve hot with parsley.

Nutrition:

- Calories: 178
- Fat: 7,9g
- Carbohydrates: 9,3g
- Protein: 5,7g

Appetizing Chana Masala

It is unusual meal but you will like it. The unique ingredients are very simple in preparing.

Prep time: 10 minutes

Cooking time: 30 minutes

Servings: 4

Ingredients:

- 8 tablespoons of mustard pasta
- ½ teaspoon of salt
- ½ teaspoon of pepper
- 1 tablespoon of garam masala
- ½ teaspoon of turmeric
- ½ teaspoon of coriander
- 8 tomatoes
- 3 tablespoon of garlic pasta
- 3 tablespoons of ginger pasta
- 5 onions
- 1 lb of kabuli chana
- 2 tablespoons of oil

Directions:

1. Sprinkle the Air Fryer with oil.
2. Then preheat it for 5 minutes to 300F.
3. Blend mustard pasta, salt, pepper, garam masala, turmeric, coriander, garlic pasta and ginger pasta in the bowl.
4. The chop tomatoes in the pieces.
5. Cut onions in the rings and add to tomatoes.
6. After that cook the kabuli chana for 15 minutes in the Air Fryer.
7. Then add all spices, tomatoes and onions and blend them together.
8. Cook for 15 minutes more.
9. Serve hot with fresh vegetables.

Nutrition:

- Calories: 189
- Fat: 2,3g
- Carbohydrates: 8,5g
- Protein: 3,1g

Appetizing and Smoked Mince

It is uncomplicated and simple meal. Enjoy with this main dish.

Prep time: 30 minutes

Cooking time: 45 minutes

Servings: 4

Ingredients:

- 2 tablespoons of oil
- ½ teaspoon of salt
- ½ teaspoon of pepper
- ½ teaspoon of red chili pepper
- 1 teaspoon of cumin
- 1 tablespoon of fresh coriander
- 1 ginger
- 1 garlic
- 4 onions
- 1 cup of yogurt
- 10 tablespoons of ghee
- 2 lbs of minced meat

Directions:

1. Place meat in the bowl.
2. Sprinkle the Air Fryer with oil.
3. Then preheat it to 350F for 5 minutes.
4. Place salt, pepper, cumin, red chili pepper, coriander, chopped ginger, garlic and ghee to the bowl with meat.
5. After that blend the ingredients.
6. Cook meat in the Air Fryer for 20 minutes.
7. Then chop onions in the pieces and add to meat.
8. Blend everything well and cook for 15 minutes.
9. Then cover the combination with yogurt and cook for 10 minutes.
10. Serve warm with basil leaves.
11. Enjoy with this meal together with your family.

Nutrition:

- Calories: 146
- Fat: 3,5g
- Carbohydrates: 8,4g
- Protein: 2,1g

<u>Tasty Potatoes with Soft Yogurt</u>

This meal is very delicious and tasty. The unique combination of potatoes and yogurt is very delicious.

Prep time: 30 minutes

Cooking time: 45 minutes

Servings: 4

Ingredients:

- ½ teaspoon of pepper
- ½ teaspoon of salt
- 2 tablespoons of oil
- 2 tablespoons of mustard oil
- 1 tablespoon of besan
- 1 coriander
- 2 curry leaves
- 1 tablespoon of turmeric
- ½ teaspoon of red chili pepper
- 2 green chili peppers
- 1 red pepper
- 1 tablespoon of cumin
- 1 lb of potatoes
- 1 garlic
- 1 cup of yogurt

Directions:

1. Mix pepper, salt, mustard oil, besan, coriander, turmeric, red chili peppers, cumin and garlic in the bowl.
2. Wash and clean potatoes.
3. Chop them in the pieces.
4. Mix potatoes with spices.
5. Sprinkle the Air Fryer with oil.
6. Preheat it to 350F.
7. Then cook potatoes for 20 minutes.
8. Chop red pepper, chili peppers in the pieces.
9. Add them to potatoes and cook for 20 minutes,
10. Cover everything with yogurt and cook for 5 minutes more.
11. Serve hot with parsley.

Nutrition:

- Calories: 123
- Fat: 3,2g
- Carbohydrates: 8,5g
- Protein: 1,2g

Piquant Mutton

If you like to eat spicy food, then this meal is exactly for you. Cook and enjoy.

Prep time: 10 minutes

Cooking time: 30 minutes

Servings: 4

Ingredients:

- 2 lbs of mutton
- ½ teaspoon of salt
- ½ teaspoon of pepper
- 2 teaspoons of oil
- 2 tablespoons of lemon juice
- 1 tablespoon of almonds
- 3 tablespoons of poppy seeds
- 1 tablespoon of garam masala
- 1/3 tablespoon of turmeric
- 2 red chili peppers
- 2 green chili peppers
- 2 cloves of garlic
- 1 cup of yogurt
- 4 onions

Directions:

1. Sprinkle the Air Fryer with oil.
2. Then preheat the Air Fryer for 5 minutes to 350F.
3. Wash and clean mutton.
4. Then chop it in the small pieces.
5. Take the bowl and blend salt, pepper, lemon juice, almonds, garam masala, turmeric, red chili peppers, green chili peppers and garlic.
6. Rub meat with spices.
7. Then cook mutton in the Air Fryer for 10 minutes.
8. Chop onions, chili peppers and add them to meat.
9. After that, cover meat with yogurt.
10. Cook for 15 minutes more.
11. Then blend everything well and cook for 5 minutes.
12. Serve hot with vegetables and dill.

Nutrition:

- Calories: 145
- Fat: 3,3g
- Carbohydrates: 9,3g
- Protein: 6,2g

Delicious Chicken with Spices

It is very easy and simple meal. When you try it, you will be very glad to prepare it again.

Prep time: 15 minutes
Cooking time: 45 minutes
Servings: 4

Ingredients:

- 1 fresh coriander
- ½ teaspoon of salt
- ½ teaspoon of pepper
- 2 tablespoons of oil
- 6 tablespoons of ghee
- ½ tablespoon of turmeric
- 10-12 peppercorns
- 4 cardamoms
- 1 tablespoon of cumin seeds
- 2 teaspoons of red chili pepper
- 2 onions
- 1 teaspoon of garlic
- 1 teaspoon of ginger
- 2 lbs of chicken

Directions:

1. Sprinkle the Air Fryer with oil.
2. Then preheat it for 5 minutes to 350F.
3. Wash and clean chicken.
4. Chop it in the pieces.
5. After that rub meat with coriander, salt, pepper, ghee, turmeric, chopped peppercorns, cumin seeds, garlic, ginger.
6. Chop onions in the pieces and add red chili pepper.
7. Mix everything well.
8. Cook for 20 minutes.
9. Then mix everything again and cook for 25 minutes more.
10. Serve hot with vegetables.

Nutrition:

- Calories: 146
- Fat: 2,1g
- Carbohydrates: 6,3g
- Protein: 5,6g

Tasty Curry with Coconut Milk

It is unusual and very delicious meal. Just cook it and you will see, how unique and tasty it is.

Prep time: 25 minutes

Cooking time: 35 minutes

Servings: 4

Ingredients:

- ½ teaspoon of pepper
- ½ teaspoon of salt
- 2 teaspoons of oil
- 1 tablespoon of coriander
- 1 teaspoon of turmeric
- 1 tablespoon of pasta
- 1 tablespoon of mustard seeds
- 2 cups of chicken stock
- 1 teaspoon of red chili pepper
- 1 ginger
- 3 green chili peppers
- 1 tablespoon of curry
- 2 cups of coconut milk
- 9 oz of tomatoes
- 1 tablespoon of garlic
- 2 lbs of shrimps
- 3 onions

Directions:

1. Sprinkle the Air Fryer with oil.
2. Preheat it for 5 minutes to 350F.
3. Put pepper, salt, coriander, turmeric, pasta, mustard seeds, red chili pepper, ginger and garlic in the Air Fryer.
4. Cook for 5 minutes.
5. Then chop green chili peppers and cook for 5 minutes more.
6. Chop tomatoes and add them to the products.
7. Cook for 10 minutes more.
8. Then add chopped onion, shrimps, coconut milk and chicken stock.
9. Cook everything for 15 minutes.
10. Serve hot with vegetables.

Nutrition:

- Calories: 156
- Fat: 2,6g
- Carbohydrates: 8,9g
- Protein: 4,5g

Flavorsome Meat with Spices

This meat is really wonderful and if you like spices, it will be great. Cook and enjoy with it.

Prep time: 30 minutes

Cooking time: 40 minutes

Servings: 4

Ingredients:

- ½ teaspoon of pepper
- ½ teaspoon of salt
- 2 tablespoons of oil
- 2-3 dry red chili peppers
- 2 black cardamoms
- 1 cup of garlic oil
- ½ cup of curd
- 2 tablespoons of ginger
- 2 tablespoons of cumin
- 3 teaspoons of garlic
- 3 tablespoons of coriander seeds
- 4 onions
- 2 lbs of mutton

Directions:

1. Sprinkle the Air Fryer with oil.
2. Preheat it for 3 minutes to 350F.
3. Wash and clean meat.
4. Then chop it in the pieces.
5. Place meat in the Air Fryer and cook for 10 minutes.
6. After that add pepper, salt, cardamom, garlic oil, ginger, cumin, chopped garlic, coriander seeds and mix everything.
7. Then chop tomatoes in the pieces, add the curd and mix everything well.
8. Cook the products for 10 minutes.
9. Chop onions in the rings and decorate the food.
10. Add also the red chili peppers.
11. Then cook everything for 20 minutes.
12. Serve hot with salad.

Nutrition:

- Calories: 189
- Fat: 5,2g
- Carbohydrates: 8,7g
- Protein: 4,9g

Sharp Chicken with Vegetables

This piquant chicken meal is appetizing. Make and eat it.

Prep time: 30 minutes

Cooking time: 30 minutes

Servings: 4

Ingredients:

- 1 lemon
- ½ teaspoon of pepper
- ½ teaspoon of salt
- 2 tablespoons of oil
- 2 green chili peppers
- 3 tablespoons of flour
- 5 oz of butter
- ½ tablespoon of garam masala
- ½ teaspoon of turmeric
- 1 teaspoon of red chili pepper
- ½ cup of yogurt
- 3 onions
- 1 tablespoon of garlic
- 2 lbs of chicken fillet

Directions:

1. Sprinkle the Air Fryer with oil.
2. Preheat the Air Fryer for 5 minutes to 350F.
3. Then slice meat in the pieces.
4. Then chop onion in the rings and blend with meat.
5. Cook in the Air Fryer for 10 minutes.
6. Then add lemon juice, pepper, salt, flour, butter, garam masala, turmeric, red chili pepper and garlic.
7. Blend everything well.
8. Cook appetizing meal for 10 minutes more.
9. Then cover cooked chicken with yogurt.
10. Cook for 5 minutes more.
11. Serve it hot with basil leaves or sprigs of mint.

Nutrition:

- Calories: 178
- Fat: 5,3g
- Carbohydrates: 8,6g
- Protein: 4,7g

Appetizing and Piquant Chicken Meat

The different combination of spices will be incredible scrumptious and delicious. Check it on your end, prepare this meat.

Prep time: 20 minutes

Cooking time: 40 minutes

Servings: 4

Ingredients:

- ½ cup of flour
- ½ teaspoon of pepper
- 2 tablespoons of oil
- ½ teaspoon of salt
- 1 pinch of saffron
- ½ teaspoon of red chili pepper
- ½ teaspoon of turmeric
- 2 tablespoons of coriander
- 10-12 peppercorns
- 12-15 cashews
- 18-20 almonds
- 4 tomatoes
- 1 cup of yogurt
- 2 onions
- 2 tablespoons of ginger
- 2 teaspoons of garlic
- 1 lb of chicken

Directions:

1. Wash and clean chicken.
2. Cut it in the pieces.
3. Then rub with salt, pepper, saffron, red chili pepper, turmeric, coriander and garlic.
4. Cook meat for 13 minutes at 350F.
5. Then chop onions and tomatoes.
6. Blend with meat and cook for 17 minutes at 330F.
7. Then add peppercorns, yogurt and blend everything well.
8. Cook for 10 minutes more.

Nutrition:

- Calories: 169
- Fat: 5,3g
- Carbohydrates: 8,4g
- Protein: 4,8g

Chicken with Tomatoes and Spices

The great meal for the family. Flavorsome and appetizing meal.

Prep time: 20 minutes

Cooking time: 40 minutes

Servings: 4

Ingredients:

- 1 lb of chicken
- 1 curry leaf
- ½ teaspoon of pepper
- ½ teaspoon of salt
- 2 tablespoons of oil
- 3 onions
- 4 tomatoes
- ½ teaspoon of turmeric
- 1 teaspoon of red chili pepper
- 6-8 red peppers
- 1,5 cup of coconut
- 1 teaspoon of garlic
- 1 teaspoon of ginger
- 2 teaspoons of lemon juice

Directions:

1. Rinse and clean chicken.
2. Cut meat in the pieces.
3. Rub it with pepper, salt, turmeric, red chili pepper, coconut, garlic, ginger and lemon juice.
4. Chop tomatoes and onions in the pieces.
5. Add them to chicken.
6. Then cut peppers and blend everything well.
7. Grease the Air Fryer with oil.
8. Cook the meal in the Air Fryer at 350F for 20 minutes.
9. Then shake everything and cook for 20 minutes more.
10. Serve hot with vegetables, parsley and sauce.

Nutrition:

- Calories: 157
- Fat: 5,7g
- Carbohydrates: 8,3g
- Protein: 4,9g

Snacks And Appetizers

Bread Rolls with Potatoes

If you have unexpected guests, then these snacks will be a good idea. They are easy and delicious and you will be delighted with the result.

Prep time: 10 minutes

Cooking time: 12 minutes

Servings:4

Ingredients:

- ½ teaspoon of salt
- ½ teaspoon of pepper
- 5 potatoes
- 2 sprigs of curry
- ½ tablespoon of mustard seeds
- ½ tablespoon of turmeric
- 1 teaspoon of coriander
- 2 onions
- 2 green chili peppers
- 8 pieces of bread
- 2 tablespoons of oil

Directions:

1. Cook potatoes and after that mix them with the spoon.
2. Then add salt, pepper, mustard seeds, coriander, turmeric and mix everything well.
3. Chop onion, add to the bowl with potatoes.
4. Then cut the green chili pepper in small pieces.
5. Mix everything again.
6. After that put bread in water and them divide the potato mixture in 8 parts.
7. Put the mixture in bread and make the rolls.
8. Sprinkle the Air Fryer with oil.
9. Then put the rolls in the Air Fryer and cook for 12 minutes at 300F.
10. Put in on the plate and decorate with the curry leaves.
11. Serve hot with sauces.

Nutrition:

- Calories: 145
- Fat: 2,1g
- Carbohydrates: 6,8g
- Protein: 1,1g

Mutton Chops

These snacks are tasty, not complicated and flavorsome. They do not take a lot of your efforts.

Prep time: 15 minutes

Cooking time: 10-11 minutes

Servings:4

Ingredients:

- 3 tablespoons of oil
- 1 tablespoon of garam masala
- ½ teaspoon of salt
- ½ teaspoon of pepper
- 1 tablespoon of ginger
- 1 tablespoon of garlic
- 3 tablespoon of red chili pepper
- 2 eggs
- 2 cups of bread crumbs
- 2 lbs of mutton chops

Directions:

1. Sprinkle the Air Fryer with oil.
2. Then mix red chili pepper, garlic, ginger, pepper, salt and garam masala in the bowl.
3. After that rub meat with these flavors.
4. Beat eggs.
5. Take meat, place it in egg and after that in the cup of bread crumbs.
6. Cook in the Air Fryer for 5-6 minutes at 300F.
7. Then put meat on the other side. Cook for 5 minutes more.
8. Serve warm with salad or vegetables.
9. Enjoy with this snack.

Nutrition:

- Calories: 359
- Fat: 10,8g
- Carbohydrates: 9,3g
- Protein: 17,3g

Chicken Bites with Ginger and Curry

It is very juicy and appetizing snack. Your guests will like them a lot because these chicken bites are very delicious.

Prep time: 15 minutes

Cooking time: 20 minutes

Servings:4

Ingredients:

- 4 tablespoons of oil
- Coriander for decoration
- ½ teaspoon of salt
- ½ teaspoon of pepper
- ½ tablespoon of garam masala
- ¼ tablespoon of turmeric
- 1 tablespoon of rec chili pepper
- 4 sprigs of curry
- ¼ cinnamon stick
- 2 tablespoon of ginger
- 3 green cardamom
- 1 bay leaf
- 1 lb of chicken

Directions:

1. Cut chicken in the pieces.
2. Then take the bowl and mix salt, pepper, garam masala, turmeric, red chili pepper, cinnamon, bay leaf, ginger and cardamom.
3. Rub the pieces of chicken with the mixture of seasonings.
4. Sprinkle the Air Fryer with oil.
5. After that preheat it to 300F.
6. Then cook it for 10 minutes.
7. After that shake the pieces of chicken well and cook for 10 minutes more at the same temperature.
8. Serve hot with salad and sauces.

Nutrition:

- Calories: 340
- Fat: 11,2g
- Carbohydrates: 9,7g
- Protein: 17,9g

Banana Chips

This snack is very easy in preparing and your guests will be surprised to see the banana chips. They are delicious and fantastic. Just try!

Prep time: 15 minutes

Cooking time: 10 minutes

Servings:4

Ingredients:

- 4 bananas
- 1 teaspoon of salt
- 1 teaspoon of pepper
- ½ teaspoon of turmeric
- ½ teaspoon of paprika
- ½ teaspoon of chat masala
- 3 tablespoons of oil
- 2 cups of water

Directions:

1. Peel bananas and cup them in the pieces.
2. Mix salt, turmeric with water.
3. Put bananas in this water.
4. If you put them there, they will have yellow color for the long time.
5. Leave them there for 10 minutes.
6. Then put bananas and mix with 1 tablespoon of oil, chat masala, pepper and paprika.
7. Sprinkle the Air Fryer with the rest of oil.
8. Preheat it to 260F.
9. Cook them for 5 minutes.
10. Then shake well and cook for 5 minutes more.
11. Serve hot and surprise your guests.

Nutrition:

- Calories: 78
- Fat: 2,1g
- Carbohydrates: 6,7g
- Protein: 1,5g

Vegetable Rolls

It is an excellent snack - pleasant, gentle, but at the same time very hearty! Just cook and you will like it.

Prep time: 15 minutes

Cooking time: 20 minutes

Servings:4

Ingredients:

- 1 spring onion
- 2 tablespoons of oil
- 1 teaspoon of soy sauce
- ½ teaspoon of salt
- ½ teaspoon of pepper
- 1 teaspoon of sugar
- 1 teaspoon of paprika
- 1 teaspoon of garlic powder
- ½ tablespoon of capsicum
- ½ teaspoon of ginger
- 2 cups of cabbage
- 1 carrot
- 2 tablespoons of flour
- 10 spring rolls sheets

Directions:

1. Chop cabbage in the small pieces and put them in the bowl.
2. Then cut carrot and mix everything.
3. Add soy sauce, salt, pepper, sugar, paprika, garlic powder, capsicum ginger and flour.
4. Mix everything well.
5. Sprinkle the Air Fryer with oil.
6. Preheat it to 300F.
7. Put the mixture on spring roll sheets and make the rolls.
8. Put them in the Air Fryer and cook for 10 minutes at 250F.
9. Then put them on another side and cook for 10 minutes more.
10. Enjoy with basil leaves or parsley.

Nutrition:

- Calories: 137
- Fat: 2,6g
- Carbohydrates: 6,2g
- Protein: 1,1g

Tortilla Chips

If you wish to prepare easy, but at the same time delicious snack, then you found the needed recipe. The result will be fantastic.

Prep time: 15 minutes

Cooking time: 8 minutes

Servings:4

Ingredients:

- 8 corn tortillas
- ½ teaspoon of salt
- ½ teaspoon of paprika
- ½ teaspoon of red chili pepper
- 3 tablespoons of oil

Directions:

1. Sprinkle the frying basket with oil.
2. Preheat the Air Fryer to 280F.
3. Take tortillas and cut then in the slices.
4. After that take the plate and mix salt, paprika and chili pepper there.
5. Rub the pieces of tortillas with spices.
6. Then put them in the Air Fryer.
7. Cook for 5 minutes.
8. Then shake them well and cook for 3 minutes more. The temperature should be 260F.
9. Serve hot with ketchup.
10. Enjoy with the scrumptious meal.

Nutrition:

- Calories: 98
- Fat: 2,3g
- Carbohydrates: 5,4g
- Protein: 1,2g

Appetizing Tacos

It is the recipe of unusual snakes. Your family will appreciate your efforts.

Prep time: 15 minutes

Cooking time: 10 minutes

Servings:4

Ingredients:

- 1 cup of feta
- ½ teaspoon of salt
- ½ teaspoon of pepper
- 2 limes
- ½ teaspoon of onion powder
- ½ teaspoon of paprika
- ½ teaspoon of red chili pepper
- 1 tablespoon of coconut oil
- 1 onion
- 3 tablespoons of olive oil
- 8 corn tortillas

Directions:

1. Take the bowl and mix salt, pepper, onion powder, paprika, red chili pepper, coconut oil there.
2. Then cut the feta in the pieces.
3. Chop onion in the rings.
4. Add onion and feta to spices.
5. Mix everything well.
6. Divide the mixture in the parts and place on tortillas.
7. Sprinkle the air frying basket with oil.
8. Then preheat it to 300F.
9. Put the tacos in the Air Fryer and cook for 5 minutes.
10. Then, put them on the other side and cook for 5 minutes more.
11. Serve hot with vegetables.

Nutrition:

- Calories: 156
- Fat: 3,1g
- Carbohydrates: 8,6g
- Protein: 1,5g

Shrimp with Lime and Tequila

These shrimps are very delicious and easy in preparing. You can be sure that you will like them a lot.

Prep time: 15 minutes

Cooking time: 15 minutes

Servings:4

Ingredients:

- 1 lime
- 12 big shrimps
- 2 oz of tequila
- 2 tablespoons of oil
- ½ teaspoon of salt
- ½ teaspoon of pepper
- ½ teaspoon of onion powder
- ½ teaspoon of garlic powder
- 1 onion

Directions:

1. Mix 1 tablespoon of oil, salt, pepper, onion powder, tequila, garlic powder and blend all well.
2. Then chop onion in the pieces and mix again.
3. Wash and clean mint.
4. Rub them with spices.
5. Leave in marinade for 10 minutes.
6. Sprinkle the frying basket with oil.
7. Preheat the Air Fryer to 350F.
8. Then put mint in the Air Fryer.
9. Cook mint for 10 minutes.
10. After that put them on the other side, and cook for 5 minutes more.
11. Sprinkle with lime juice and serve hot.
12. You can eat them with sauces.

Nutrition:

- Calories: 136
- Fat: 2,7g
- Carbohydrates: 9,4g
- Protein: 10,2g

Garlic Shrimps with Spinach

This combination of shrimps and spinach will surprise your guests a lot. It is very delicious and amazing. You should cook it.

Prep time: 15 minutes

Cooking time: 15 minutes

Servings:4

Ingredients:

-

Directions:

1. Cook spinach and when it is ready just chop it in the small pieces.
2. After that mix salt, pepper, garlic powder, paprika and 1 tablespoon of oil.
3. Then chop onion and mix everything together.
4. Sprinkle the Air Fryer with oil.
5. Preheat it to 350F.
6. Then add the cooked shrimps and put in the Air Fryer.
7. Cook for 10 minutes at 350F.
8. Then mix everything and cook for 5 minutes more.
9. Enjoy with this snack.

Nutrition:

- Calories: 161
- Fat: 2,3g
- Carbohydrates: 9,7g
- Protein: 10,9g

Mexican Chicken

This snack is very easy and will not take a lot of your time. Also, they are very delicious. Just cook and check it on your end.

Prep time: 15 minutes

Cooking time: 25 minutes

Servings:4

Ingredients:

- ½ teaspoon of salt
- ½ teaspoon of pepper
- 3 tablespoons of lime juice
- ¼ cup of cilantro
- ¼ cup of salsa
- ½ teaspoon of ground coriander
- ½ tablespoon of cumin
- 1 cup of cooked chicken

Directions:

1. Take the bowl and mix salt, pepper, cilantro, salsa, ground coriander, cumin and lime juice together.
2. Then chop chicken in the pieces.
3. Rub them with spices.
4. After that sprinkle the Air Fryer basket with oil.
5. Then cook chicken for 20 minutes at 200F.
6. After that, when you see the crispy skin, put chicken on the other side.
7. Cook for 5 minutes more.
8. Serve hot with ketchup.

Nutrition:

- Calories: 239
- Fat: 7,2g
- Carbohydrates: 8,4g
- Protein: 10,8g

Crab Sticks

It is very interesting and flavorsome snack. If you wish to prepare something unusual for your guests, you should prepare exactly this meal.

Prep time: 15 minutes

Cooking time: 23 minutes

Servings:4

Ingredients:

- 1 packet of ready frozen crab sticks
- ½ teaspoon of pepper
- ½ teaspoon of salt
- ½ teaspoon of garlic powder
- ½ teaspoon of onion powder
- ½ teaspoon of paprika
- 2 tablespoons of oil

Directions:

1. Mix pepper, salt, garlic powder, onion powder, paprika and 1 tablespoon of oil in the container.
2. If you wish, it is possible to cut the crab sticks in the small parts.
3. Then rub them with spices.
4. After that sprinkle the basket with the rest of oil.
5. Cook the crab sticks for 20 minutes at 260F.
6. When they have golden color, put then on the other side. Cook for 3 minutes.
7. Serve warm and enjoy with your great crab snacks.

Nutrition:

- Calories: 135
- Fat: 2,1g
- Carbohydrates: 7,8g
- Protein: 10,9g

Potato Chips

These potato chips will be the best recipe for you. They are crispy and delicious.

Prep time: 5 minutes

Cooking time: 22 minutes

Servings: 3

Ingredients:

- ½ teaspoon of salt
- ½ teaspoon of pepper
- ½ teaspoon of paprika
- ½ teaspoon of red chili pepper
- 2 tablespoons of oil
- 5 big potatoes

Directions:

1. Wash and clean potatoes.
2. Then cut them in the slices.
3. After that mix salt, pepper, paprika, 1 tablespoon of oil and red chili pepper in the bowl.
4. Rub potatoes with the different spices.
5. Sprinkle the frying basket with two tablespoons of oil.
6. Place chips in the Air Fryer and cook for 12 minutes at 260F.
7. Then shake chips well and cook for 10 minutes more at 230F.
8. Serve hot meal with the different sauces or fresh vegetables.
9. Enjoy with them.

Nutrition:

- Calories: 136
- Fat: 3,4g
- Carbohydrates: 2,2g
- Protein: 1,2g

Cauliflower Bites

It is healthy and delicious snack. You will spend only 20 minutes for the preparing this meal

Prep time: 5 minutes

Cooking time: 25 minutes

Servings: 4

Ingredients:

- 2/3 cup of hot sauce
- ½ teaspoon of salt
- ½ teaspoon of pepper
- ½ teaspoon of garlic powder
- ½ teaspoon of onion powder
- 1 tablespoon of coconut oil
- 2 tablespoons of oil
- 1 big head of cauliflower

Directions:

1. Wash and clean cauliflower.
2. Then chop it in the pieces.
3. Put in the container and mix with salt, pepper, garlic powder and onion powder.
4. Blend these ingredients well.
5. Then sprinkle the Air Fryer basket with 2 tablespoons of oil.
6. Preheat it to 350F.
7. Cook cauliflower for 15 minutes.
8. Then add coconut oil, blend the components and cook for 5 minutes.
9. After that cover all products with sauce. Cook for 5 minutes more at the same temperature.
10. Enjoy with your friends or relatives.

Nutrition:

- Calories: 120
- Fat: 3,1g
- Carbohydrates: 8,9g
- Protein: 1,1g

Grilled Broccoli

This snack is fantastic. Broccoli with spices is very delicious. Your guests will be surprised with this meal a lot.

Prep time: 5 minutes

Cooking time: 10 minutes

Servings: 4

Ingredients:

- 4 cups of broccoli
- 2 teaspoons of garlic powder
- 1 teaspoon of pepper
- ½ teaspoon of salt
- 1/8 teaspoon of paprika
- 1/6 teaspoon of oregano
- 1 tablespoon of olive oil
- 1 tablespoon of coconut oil
- 1 big red pepper
- ½ teaspoon of onion powder
- ½ cup of sauce

Directions:

1. Wash and cup broccoli in the pieces.
2. Wash and chop the red pepper in the strings.
3. Mix the red pepper with broccoli.
4. Then add pepper, salt, paprika, oregano, coconut oil and onion powder to the bowl with vegetables.
5. Mix everything well.
6. Sprinkle the crying basket with oil.
7. Preheat the Air Fryer to 350F.
8. Cook broccoli for 5 minutes at 350F.
9. Then shake well, cover with sauce and cook for 5 minutes again.
10. Serve warm and you can decorate it with parsley or sprinkle with lemon.

Nutrition:

- Calories: 89
- Fat: 2,1g
- Carbohydrates: 10,1g
- Protein: 1,2g

Cauliflower with Turmeric and Garlic

You will like this cauliflower because of the different spices. This snack is very delicious and tasty. Do not miss your chance to cook it.

Prep time: 5 minutes

Cooking time: 30 minutes

Servings: 4

Ingredients:

- 3 tablespoons of chopped cilantro
- 3 scallions
- 1 cauliflower
- ½ teaspoon of salt
- ½ teaspoon of pepper
- ½ teaspoon of ground cumin
- 1 teaspoon of ground turmeric
- 4 garlic cloves
- 2 tablespoons of lemon juice
- 3 tablespoons of oil

Directions:

1. Preheat the Air Fryer to 350F.
2. Sprinkle it with oil.
3. Then wash and chop cauliflower in the small pieces.
4. Add salt, pepper, scallions, ground cumin, ground turmeric, chopped garlic cloves and mix the components.
5. Then place in the Air Fryer and cook at 350F for 20 minutes.
6. Then mix cauliflower and add 2 tablespoons of lemon juice.
7. After that cook for 10 minutes at 300F.
8. Serve hot.

Nutrition:

- Calories: 110
- Fat: 1,6g
- Carbohydrates: 8,2g
- Protein: 1,2g

Fish and Seafood

Air Fryer Fish Chips

Delicious chips for the big company. Just try and enjoy.

Prep time:10 minutes

Cooking time:15 minutes

Servings:4

Ingredients:

- 1 tablespoon of parsley
- Salt
- Pepper
- 1 egg
- 2 fish fillets
- 1 oz of chips
- 1 big lemon
- 3 tablespoons of bread crumbs

Directions:

1. Cut every fillet into two pieces. You should have 4 parts of fish fillet.
2. Mix lemon juice with seasonings and put aside.
3. Put in the food processor salt, pepper, parsley, rest of lemon, chips and bread crumbs. Then just grind them.
4. Put the mixture in the backing basket for the Air Fryer.
5. Fish should be placed in beaten egg firstly and covered with it.
6. Then put fish in the mixture of bread crumbs and seasonings.
7. The process of cooking takes 15 minutes and the temperature should be 370F.
8. Enjoy with the delicious meal.

Nutrition:

- Calories: 830
- Fat: 13,1g
- Carbohydrates: 125,1g
- Protein: 43,2g

Lemon Fish

If you like fish, then this recipe is the best choice. Fish will be the main dish on your table after reading it.

Prep time:0-5 minutes

Cooking time: 31minutes

Servings:4

Ingredients:

- Lettuce 2-3 leaves
- 1 teaspoon of red chili sauce
- 4 teaspoons of corn flour slurry
- 1 egg (white)
- Juice of lemon
- 2 teaspoons of oil
- 2 teaspoons of green chili sauce
- Salt
- ¼ cup sugar
- 1 big lemon

Directions:

1. Chop lemon. Put it in the bowl.
2. Boil half of cup of water and add sugar.
3. Stir it till sugar dissolves completely.
4. Take the deep bowl.
5. Put egg white, 2 tablespoons of oil, salt, 1 cup of flour, green chili sauce in the bowl.
6. Mix the ingredients well.
7. Then add 3 tablespoons of water and whisk it.
8. Put fillets in butter and after that cover with the refined flour.
9. Brush the basket with oil and also cook the Air Fryer.
10. Put fillets in the basket for Air Fryer and cook for 15-20 minutes. The temperature should be 365F.
11. Add salt to sugar. Mix it well in the pan. Then add red chili sauce, corn flour slurry, lemon and lemon juice and mix everything.
12. Put the leaves of salad on the plate.
13. After that, put fish on the leaves and pour lemon sauce over it.
14. You can be sure, that you will like it.

Nutrition:

- Calories: 972
- Fat: 15,6g
- Carbohydrates: 155,1g
- Protein: 49,5g

Fish and Chips with Tartar Sauce

Do you want to find the easy and at the same time simple fish meal? You have just found it.

Prep time: 30 minutes

Cooking time: 40 minutes

Servings:3

Ingredients:

Tartar sauce:

- Salt and black pepper
- 1 lemon (juice)
- 1 little shallot
- 2 tablespoons of jalapenos
- 3 tablespoons of capers
- 1 cup of mayonnaise

Chips:

- 2 pieces of garlic
- 1 tablespoon of oil
- 1 teaspoon of rosemary
- 2 big potatoes

Fish:

- 1 teaspoon of salt
- 1 teaspoon of pepper
- 1 tablespoon of oil
- 1 tablespoon of parsley
- ½ cup bread crumbs
- ½ cup flour
- 2 eggs
- 15 oz of white fish fillet

Directions:

1. The Air Fryer should be preheated to 392F for 5 minutes.
2. Take the bowl and mix salt, pepper, oil, parsley and bread crumbs.
3. Fish should be chopped into 8 pieces.
4. Put one piece of fish in flour.
5. After that, put fish into beaten egg and after that in bread crumbs.
6. Put fish in the Air Fryer. Cook up to 15 minutes.
7. Cut potatoes for chips, chop them into slices and put in the bowl with salted water for 20 minutes.
8. Preheat the Air Fryer up to 360F for 5 minutes.
9. Mix the ingredients and put them in Air Fryer for 20 minutes.
10. For sauce just mix the ingredients and leave them for 20 minutes.

Nutrition:

- Calories: 856
- Fat: 16g
- Carbohydrates: 159,7g
- Protein: 44,9g

Broiled Tilapia

This recipe will surprise you a lot. You will not spend too much time on preparing it and it is very delicious.

Prep time: 10 minutes
Cooking time: 7 minutes
Servings:2

Ingredients:

- Tilapia fillets 1 lb
- Lemon pepper
- Salt
- 1 tablespoon of oil

Directions:

1. If the pieces of fillet are frozen, thaw them.
2. Spray the basket for your Air Fryer with oil. Also, it is possible to use the cooking spray.
3. Preheat the Air Fryer for 5 minutes to the temperature 392F.
4. Put fish in the basket.
5. Set the temperature 360F and cook it for 7 minutes.
6. Fish should have golden color. If you have too big pieces of fish, cook it a bit longer, for example 2-3 minutes more.
7. It is possible to add your favorite vegetables like tomato, cucumber or something like this to fish.
8. Your family will enjoy with it.

Nutrition:

- Calories: 900
- Fat: 17g
- Carbohydrates: 160g
- Protein: 43g

Cod Fish with Crispy Skin

This is very healthy food and it is really worth to taste. It will be great meal for dinner.

Prep time: 2 minutes

Cooking time: 7 minutes

Servings:4

Ingredients:

- Coriander
- Green onion
- White onion
- 5 pieces of ginger
- 3 tablespoons of oil
- 5 teaspoons of sugar
- 5 teaspoons of soy sauce
- 1 cup of water
- Sesame oil
- Salt and sugar
- 15 oz of cod fish

Directions:

1. Wash fish and after that dry it.
2. Add salt, 3 teaspoons of sugar and oil to fish and leave for 15 minutes.
3. Preheat the Air Fryer to 350F for 3 minutes.
4. Cook fish during 12 minutes.
5. Then prepare sauce.
6. Pour 1 cup of water in the pan and boil.
7. Add sugar, soy sauce and wait up to 5 minutes.
8. Cook oil in the small pen, add onion and ginger. It will take 10 minutes.
9. Take fish from Air Fryer and out on the dish. Pour with sauce and oil.
10. You can cook it any time you wish, it is really delicious.

Nutrition:

- Calories: 890
- Fat: 16g
- Carbohydrates: 170g
- Protein: 45g

Tasty Fish

Prepare it quickly and delicious. It will not take too much of your time and money.

Prep time: 10 minutes

Cooking time: 12 minutes

Servings:4

Ingredients:

- 1 lemon
- 1 egg
- 4 tablespoons of oil
- 4 fish fillets
- 15 oz of bread crumbs

Directions:

1. Put fish and wash it.
2. Preheat the Air Fryer up to 370F for 4-5 minutes.
3. Mix oil and bread crumbs together.
4. Put the piece of fish fillet into beaten egg.
5. Then put it in the bowl with bread crumbs.
6. Check if fish is covered with bread crumbs completely.
7. Put fish in the Air Fryer and cook it for 12 minutes. The temperature of the Air Fryer should be no more than 360F.
8. Take fish out of Air Fryer and sprinkle it with lemon.
9. Serve it hot, because it is the most delicious when it has been just prepared.
10. Just enjoy with this fish.

Nutrition:

- Calories: 883
- Fat: 17,5g
- Carbohydrates: 169g
- Protein: 49g

Herb and Garlic Fish Fingers

This recipe can be used by everyone. Try to prepare fish and you will get the perfect dish in the result.

Prep time: 10 minutes

Cooking time: 30 minutes

Servings:4

Ingredients:

- Oil
- 1 cup of bread crumbs
- ¼ teaspoon of backing soda
- 2 eggs
- 2 tablespoons of lemon juice
- 2 tablespoons of cornflour
- 2 tablespoons of maida
- 1 teaspoon of garlic
- ½ teaspoon of pepper
- 2 teaspoons of mixed herbs
- ½ teaspoon of red chili pepper
- ½ teaspoon of turmeric
- ½ teaspoon of salt
- 10 oz of seer fish

Directions:

1. Wash fish and put in the bowl.
2. Add salt, pepper, red chili pepper, lemon juice, herbs, garlic and mix everything.
3. Put aside for 10 minutes.
4. Mix the corn flour, maida, egg and soda in another bowl.
5. Add fish from the first bowl in this one and leave for another 10 minutes.
6. Mix herbs and garlic and cover fish fingers with them.
7. Preheat the Air Fryer to 370F for 2-3 minutes.
8. Put the foil on the backing basket.
9. Then sprinkle the foil with oil and put fish fingers.
10. Cook up to 10 minutes.

Nutrition:

- Calories: 869
- Fat: 19,1g
- Carbohydrates: 130g
- Protein: 46,8g

Grilled Fish Fillet with Pesto Sauce

You can use this recipe of fish every day. It will be great as for the usual meal as for your guests.

Prep time: 10 minutes

Cooking time: 8 minutes

Servings:3

Ingredients:

- 1 cup of olive oil
- 1 tablespoon of parmesan cheese
- 2 pieces of garlic
- 2 tablespoons of pinenuts
- 1 bunch of basil
- Pepper and salt
- 3 big pieces of white fish fillets

Directions:

1. Firstly, preheat the Air Fryer tO 365F.
2. Then wash fish.
3. Cover it with oil and add salt and pepper.
4. Put it in the cooking basket and then put in the Air Fryer.
5. Cook it for 8 minutes.
6. Take basil and grind with cheese, garlic, pinenuts in the food processor.
7. Add salt.
8. Put fish on the dish and cover with sauce.
9. There is no need to serve it hot, so it is possible to wait 10-15 minutes.
10. This meal will be your favorite one.

Nutrition:

- Calories: 785
- Fat: 8g
- Carbohydrates: 120g
- Protein: 47g

Fish and Chips

You can prepare this meal even if you are too busy. It is very easy and delicious.

Prep time: 40 minutes

Cooking time: 12 minutes

Servings:4

Ingredients:

- ½ tablespoon of lemon juice
- 1 tablespoon of oil
- 10 oz of red potatoes
- 1 big egg
- 1 oz of chips
- 6 oz of white fish fillet

Directions:

1. The Air Fryer should be preheated to 370F for 2-3 minutes.
2. Fish should be divided into 4 pieces.
3. Then rub it with salt, pepper and lemon juice.
4. Leave it for 5 minutes.
5. Grind chips with the help of food processor.
6. Beat egg into the bowl.
7. Put the piece of fish in egg firstly and then into the pieces of chips.
8. Clean potatoes and divide into long pieces.
9. Leave them in water for 25-30 minutes.
10. After that dry the potato and put in oil.
11. Boil potatoes for 5 minutes.
12. After that, the separator should be inserted in the Air Fryer.
13. Put potatoes on one side and fish on another one.
14. Cook for 12 minutes at 360F.
15. It is very delicious, you will like it.

Nutrition:

- Calories: 691
- Fat: 12g
- Carbohydrates: g
- Protein: 41g

Fish Sticks

The easiest and very cheap snack from fish. You can enjoy it any time you wish, because it will not take a lot of efforts to be prepared.

Prep time: 10-15 minutes

Cooking time: 12 minutes

Servings:4

Ingredients:

- 3 tablespoons of milk
- 1 cup of flour
- ½ teaspoon of black pepper
- ¼ of teaspoon of sea salt
- 1 cup Panko
- 1 big egg
- 1 pound of cod

Directions:

1. Mix milk and egg in the bowl.
2. Put bread crumbs on the bottom of the pan.
3. Add seasonings to milk and egg.
4. Then mix everything well.
5. Put fish and cover it with bread crumbs.
6. After that put the dish in the fry basket.
7. Pour fish with the mixture of egg, milk and seasonings.
8. Cook it for 12 minutes at 390F.
9. Then take it out of Air Fryer and leave for 5-10 minutes, because it will be very hot.
10. The result will exceed your expectations.

Nutrition:

- Calories: 737
- Fat: 15,9g
- Carbohydrates: 121g
- Protein: 46,8g

Fish Chili Basil

The interesting combinations of ingredients are hidden in this recipe. When you try it, you will not be able to refuse from this fish anymore.

Prep time: 0-5 minutes

Cooking time: 16-20 minutes

Servings: 4

Ingredients:

- 2 teaspoons of red chili pepper
- 2 tablespoons of soy sauce
- 1 tablespoon of garlic
- 1 tablespoon of oil
- Salt
- Pepper
- ¼ cup and 1 tablespoon of cornstarch
- 6-7 red chilies
- 15 oz of fish fillet

Directions:

1. Mix fish with salt, pepper and 1 cup of cornstarch
2. Preheat the Air Fryer for 3 minutes to 370F
3. Put fish into the basket and cook for 8 minutes. Fish should have golden color.
4. Cook oil (1 tablespoon) and mix it with garlic and red chili, which should be chopped.
5. After that add water to the mixture.
6. Then add chili flakes and soy sauce.
7. Put 1 tablespoon of cornstarch and wait. Sauce should be thickened.
8. Pour fish with sauce.

Nutrition:

- Calories: 1157
- Fat: 89,3g
- Carbohydrates: 127g
- Protein: 60,4g

Cajun Fish

Do you have some unexpected guests? Do not worry, this recipe can help in these situations, because it needs only a few minutes of your time.

Prep time: 0-5 minutes

Cooking time: 6-10 minutes

Servings: 4

Ingredients:

- 1 tablespoon of lemon juice
- 2 pieces of fish fillets
- 2 tablespoons of refined flour
- 1 tablespoon of red chili flakes
- ½ tablespoon of garlic
- Salt
- 1 teaspoon of stock cubes
- 1 teaspoon of herbs
- 1 teaspoon of red chili powder
- 2 tablespoons of oil

Directions:

1. Grind salt, garlic, herbs, chili powder, stock cubes and chili flakes in the food processor.
2. After that add the refined flour and mix.
3. Divide fillet into the pieces.
4. Rub these pieces with lemon juice.
5. Preheat the Air Fryer. It is required 2 minutes to392F.
6. Then put fish into sauce and after that cook it in the Air Fryer for 7 minutes. The temperature should be 370F.
7. You will see the crispy skin on fish.
8. The meal is really tasty.

Nutrition:

- Calories: 1011
- Fat: 73,9g
- Carbohydrates: 26g
- Protein: 60,3g

Steamed Basa Fish

According to this recipe you will get incredibly juicy, delicate and delicious fish. Just try and you will see it.

Prep time: 11-15 minutes

Cooking time: 11-15 minutes

Servings: 4

Ingredients:

- Salt
- 1 tablespoon of lemon juice
- 8-10 sprigs of coriander
- 1 teaspoon of cumin
- 1 green chili pepper
- ½ tablespoon of garlic
- ¼ cup coconut
- 2-3 leaves of basil
- 1 tablespoon of black pepper
- 1 tablespoon of oil
- 1 tablespoon of herbs
- 15 oz of Basa fish fillets

Directions:

1. Wash and cut fish into the pieces.
2. Rub it with salt and black pepper.
3. Preheat the Air Fryer for 2 minutes under 350F.
4. Put fish into the backing basket, but firstly put the foil and sprinkle it with 1 tablespoon of oil.
5. Mix the seasoning in the bowl.
6. After that put everything from the bowl into the backing basket on fish and mix it.
7. Cook for 11-15 minutes under 390F. Fish should have brown color.
8. You will cook this meal very often, because it is delicious.

Nutrition:

- Calories: 1025
- Fat: 78g
- Carbohydrates: 29g
- Protein: 63,5g

Spinach Fish Rolls

You should not visit restaurant to try this meal. It is easy to prepare these rolls at home.

Prep time: 0-5 minutes

Cooking time: 8-10 minutes

Servings: 4

Ingredients:

For sauce:

- 1 teaspoon of lemon juice
- 3-4 chopped chives
- 1 tablespoon of oil

For fish:

- 1 teaspoon of red chili flakes
- 1 cup of cheese (paneer)
- 2 teaspoons of lemon juice
- 2 teaspoons of garlic pasta
- 1 teaspoon of mustard pasta
- 1 teaspoon of black pepper
- Salt
- 8-10 Spinach leaves

Directions:

1. Preheat the Air Fryer to 392F for 5 minutes.
2. Wash fish.
3. Chop it into pieces and put in the bowl.
4. Add lemon juice, pepper, garlic pasta, mustard pasta and mix them together.
5. Put aside for 15 minutes.
6. After that put cheese in another bowl.
7. Add red chili flakes and salt.
8. Put fish on the plate. Then put spinach on it and after that put the mixture of cheese.
9. Roll everything tightly.
10. Put them in the Air Fryer and cook for 8-10 minutes at 370F.
11. Your family will appreciate your efforts.

Nutrition:

- Calories: 1080
- Fat: 72g
- Carbohydrates: 49g
- Protein: 69,9g

Lemon Chili Fish with Kurmura

It is very easy in preparing, but you even cannot imagine how delicious this fish is. Prepare and enjoy.

Prep time: 6-10 minutes

Cooking time: 16-25 minutes

Servings: 4

Ingredients:

- 2-3 teaspoons of castor sugar
- 1 teaspoon of coriander
- 1 red capsicum
- 15-18 curry leaves
- 1 teaspoon of red chili pepper
- Salt
- ¼ teaspoon of turmeric
- 1 onion
- 3-4 green chilies
- 1 lemon juice
- 2 cups of puffed rice
- 2 pieces of fish fillets

Directions:

1. Wash fish and chop it.
2. Chop onion in the smallest pieces.
3. Add there chopped green chili.
4. Put rice in the Air Fryer and cook it with the cup of water for 10 minutes at 350F.
5. Take rice out of Air Fryer into the bowl.
6. Add the turmeric and red chili pepper.
7. Add the green chili, curry and onion.
8. After that chop red capsicum and add coriander with lemon juice.
9. Put everything in the backing basket and add fish,
10. Cook for 10 minutes at 360F.
11. Just cook and enjoy.

Nutrition:

- Calories: 966
- Fat: 72g
- Carbohydrates: 64,3g
- Protein: 36,9g

Chicken Nuggets

This meal will decorate every dish. It is very tasty and does not take too much time for preparing.

Prep time:10 minutes

Cooking time: 10-15 minutes

Servings:4

Ingredients:

- Salt
- Pepper
- 1 tablespoon of parsley
- 1 tablespoon of paprika
- 1 tablespoon of oil
- 2 beaten eggs
- 1 tablespoon of ketchup
- 1 tablespoon of garlic puree
- 1 lb of chicken breast
- 2 pieces of wholemeal bread (bread crumbs)

Directions:

1. Mix bread crumbs with paprika, pepper and salt.
2. After that add oil and mix everything well.
3. You should get the batter.
4. Grid chicken in the food processor and then add ketchup, one beaten egg, garlic and parsley.
5. Put the second beaten egg into the bowl.
6. Create cutlets and put them in chicken nuggets and after cover completely with egg.
7. Cook for 10-15 minutes at 392F.
8. Serve hot and enjoy with this dish.

Nutrition:

- Calories: 332
- Fat: 13,7g
- Carbohydrates: 7g
- Protein: 38g

Chicken Kievs

Incredibly delicious pieces of chicken. You even cannot imagine, that it can be so tasty.

Prep time:10 minutes

Cooking time: 25 minutes

Servings:2

Ingredients:

- Salt
- Pepper
- 1 big egg
- 1 tablespoon of parsley
- ¼ tablespoon if garlic puree
- 4 oz of soft cheese (the flavor of herb and garlic)
- 1 large chicken breast
- Bread crumbs

Directions:

1. Put in the bowl cheese, half of parsley and garlic.
2. Mix everything well.
3. Divide chicken into two pieces.
4. Put the mixture of garlic, parsley and cheese between two pieces of chicken.
5. Then take the bowl and put salt, pepper, parsley and bread crumbs there
6. Beat eggs and cover chicken with it.
7. Then put chicken in bread crumbs.
8. Cook in the Air Fryer for 25 minutes at 356F.
9. Serve the hot meal.

Nutrition:

- Calories: 356
- Fat: 14,7g
- Carbohydrates: 6,5g
- Protein: 38,5g

Chicken Wings

If you are looking for something that can be great for the picnic – then these wings will be the great choice. Spicy and tasty, you will want more them to eat.

Prep time:5 minutes

Cooking time: 30 minutes

Servings:2

Ingredients:

- Salt
- Pepper
- 1 tablespoon of Chinese spices
- 1 tablespoon of mixed spices
- 1 tablespoon of soy sauce
- 4 chicken wings

Directions:

1. Mix salt, pepper, soy sauce and spices in the deep bowl.
2. After that put chicken wigs in spices and cover the wigs with these spices.
3. Then put the silver foil in the basket for Air Fryer.
4. Put chicken wings on the foil.
5. Add the rest or spices in the bowl.
6. Cook at 356F for 15 minutes.
7. Then put the wings on another side and cook for another 15 minutes at 392F
8. Serve hot and get the delicious chicken wings for your family.

Nutrition:

- Calories: 372
- Fat: 15,2g
- Carbohydrates: 7,1g
- Protein: 37,5g

Popcorn Chicken

Do you want to get the tasty, juicy and fragrant chicken? You will not spend too much of your time to prepare delicious chicken.

Prep time:10 minutes

Cooking time: 12 minutes

Servings:12

Ingredients:

- Pepper
- Salt
- 2 oz of flour
- 1 egg
- 3 oz of bread crumbs
- 2 teaspoon of chicken spices
- 1 chicken breast

Directions:

1. Put chicken in the food processor and blend it.
2. You should get minced chicken in the result.
3. Take the bowl and put flour there.
4. After that beat egg in another bowl.
5. Mix chicken spices, bread crumbs, salt and pepper in the third bowl.
6. Create the balls from chicken and put them in flour.
7. After that put these balls in egg and cover completely.
8. Then put them in bread crumbs with spices.
9. Put in the Air Fryer and cook for 10-12 minutes at 356F.
10. At a result, you will get the golden chicken balls with the crispy skin.

Nutrition:

- Calories: 362
- Fat: 14,8g
- Carbohydrates: 6,7g
- Protein: 36,8g

Chicken Burger

If you wish to have some simple, but at the same time delicious snack with chicken – then you should prepare this chicken burger. You will be really surprised with this recipe.

Prep time:10 minutes

Cooking time: 15 minutes

Servings:4

Ingredients:

- 1 tablespoon of paprika
- 1 tablespoon of mustard powder
- 1 tablespoon of Worcester sauce
- ½ cup of bread crumbs
- 1 tablespoons of chicken spices
- 2 oz of flour
- 6 chicken breasts
- 1 egg
- Salt
- Pepper

Directions:

1. Blend chicken with the help of the food processor.
2. After that put chicken in the bowl.
3. Add mustard, salt, pepper, paprika and Worcester sauce.
4. Create the little burgers from chicken and leave them.
5. Beat egg in one bowl.
6. Put flour in the second one.
7. Mix bread crumbs and chicken spices in the separate bowl.
8. Then put your burgers in flour and after that in the bowl with egg.
9. When they are completely covered with egg, put them in the bowl with bread crumbs and spices.
10. After that put chicken in the Air Fryer.
11. It should be cooked for 15 minutes at 356F.
12. When chicken burgers are ready, enjoy them, because they can be the great snack.

Nutrition:

- Calories: 368
- Fat: 14,5g
- Carbohydrates: 6,2g
- Protein: 36,4g

Chicken Strips

A delicious meal from usual ingredients. Your guests will be delighted - all without exception.

Prep time:10 minutes

Cooking time: 12 minutes

Servings:8

Ingredients:

- Salt
- Pepper
- 2 tablespoons of dry coconut
- 2 tablespoons of plain oats
- 1 teaspoon of chicken spices
- 1 egg
- 2 oz of flour
- 3 oz of bread crumbs
- 1 chicken breast

Directions:

1. Take the bowl and chop chicken into small pieces.
2. Put pepper, salt, bread crumbs, coconut, oats and chicken spices in another bowl.
3. Mix everything well.
4. Then beat egg in another bowl.
5. Take the third bowl and put flour there
6. Take the piece of chicken and put it in flour, then in egg and after that in the mixture of different spices.
7. Cook in the Air Fryer at 360F for 8 minutes.
8. After that put chicken on another side and cook for 4 minutes at 320F.
9. Then enjoy with the tasty and delicious chicken at dinner table.

Nutrition:

- Calories: 364
- Fat: 14,7g
- Carbohydrates: 5,9g
- Protein: 36,9g

Tasty Backed Chicken

This meal is very easy. If you prepare it one time, you will often prepare it for different holidays.

Prep time:10 minutes

Cooking time: 18 minutes

Servings:4

Ingredients:

- 2 oz of flour
- 2 eggs
- 2 tablespoons of chicken spices
- 9 oz of bread crumbs
- 1 chicken
- 6 oz of cheese

Directions:

1. Chop chicken in the small pieces.
2. Take the bowl and mix spices with flour.
3. But bread crumbs in another bowl.
4. Beat eggs in the bowl well.
5. Take the piece of chicken and put in flour.
6. Then cover it with egg and put in bread crumbs.
7. Sprinkle the Air Fryer basket with oil.
8. Cook for 350F for 15 minutes.
9. Then add cheese and cook for 3 minutes more.
10. You will appreciate this recipe, because it is very tasty.
11. Cook this meal 1 time and you will like it.

Nutrition:

- Calories: 371
- Fat: 15,9g
- Carbohydrates: 6,5g
- Protein: 36,8g

Buffalo Chicken

This dish is suitable for a holiday table and for a very ordinary dinner. It will be appreciated as by sportsmen as by gourmets.

Prep time: 20 minutes

Cooking time: 16 minutes

Servings: 4

Ingredients:

- A cup of bread crumbs
- 1 teaspoon of hot sauce
- 1 tablespoon of hot sauce
- ¼ cup of egg substitute
- ½ cup of plain fat-free Greek yogurt
- 1 tablespoon of cayenne pepper
- 1 tablespoon of garlic pepper seasoning
- 1 tablespoon of sweet paprika
- 1 pound of skinless chicken

Directions:

1. Put one tablespoon of hot sauce and egg substitute in the bowl.
2. Then whisk yogurt and add the teaspoon of hot sauce.
3. Mix everything in the bowl.
4. After that put garlic pepper, bread crumbs, paprika and cayenne pepper in another bowl.
5. Mix everything well.
6. Then put the pieces of chicken in yogurt. Leave them for 2-3 minutes there.
7. After that put every piece of chicken in the crumbs and cover completely.
8. Then put them in the Air Fryer. Cook for 8 minutes at 370F on one side.
9. After that put them on another side and cook for another 8 minutes.
10. Enjoy with the ready chicken pieces. They should have brown color.

Nutrition:

- Calories: 376
- Fat: 15,1g
- Carbohydrates: 6,3g
- Protein: 36,6g

Peppery Chicken

This meal will be the favorite one. It is easy in preparing and too tasty.

Prep time: 10 minutes

Cooking time: 20 minutes

Servings: 4

Ingredients:

- 1 big egg
- Backing spray
- 1 teaspoon of pepper
- 1 teaspoon of salt
- 2 cups of flour
- 1 cup of buttermilk
- 3/2 pounds of chicken parts

Directions:

1. Wash chicken and put it in the bowl with buttermilk.
2. Mix salt, pepper and flour together in the bowl and add to chicken.
3. Rub chicken with the mixture of spices and four.
4. After that beat egg in the bowl and cover chicken with it. It should be completely covered with egg.
5. Sprinkle the backing basket with the backing spray.
6. Cook at 390F for 20 minutes.
7. As the result, you will get the hot chicken parts with the crispy skin.
8. You will be delighted with this meal.

Nutrition:

- Calories: 376
- Fat: 16,3g
- Carbohydrates: 6,2g
- Protein: 36,1g

Roast Chicken

This ruddy chicken has the crispy crust? This tasty meal is fantastic, you will like it.

Prep time: 10 minutes

Cooking time: 50 minutes

Servings: 4

Ingredients:

- 2 tablespoons of garlic pepper
- 2 tablespoon of cayenne pepper
- 2 tablespoons of dry mustard
- 2 tablespoons of dried thyme
- ¼ cup of brown sugar
- ¼ cup of Italian seasoning
- ¼ cup of garlic powder
- ¼ cup of onion powder
- ¼ cup of paprika
- ¾ cup of kosher salt
- 4,25 pound of chicken

Directions:

1. Wash, clean and dry chicken.
2. Then rub it with all seasonings.
3. Spray the basket for Air Fryer with the cooking spray and put chicken there.
4. Cook it in the Air Fryer for 30 minutes at 330F.
5. After that put chicken on the other side.
6. Cook it for 20 minutes more.
7. Serve it hot and your family will be glad to eat this meal.

Nutrition:

- Calories: 369
- Fat: 14,9g
- Carbohydrates: 7,1g
- Protein: 34,9g

Fried Crispy Chicken

Roast chicken is a familiar meal. However, note that this recipe contains some interesting nuances.

Prep time: 15 minutes

Cooking time: 35 minutes

Servings: 4

Ingredients:

- 2 beaten eggs
- ¼ cup of milk
- Pepper
- Salt
- 1 teaspoon of barbecue chicken spice
- 1 teaspoon of basil
- 2 teaspoons of paprika
- 2 teaspoons of garlic
- 2 teaspoons of seasoning salt
- 3 cups of flour

Directions:

1. Put 1 cup of flour in the bowl and add salt and pepper there.
2. Mix everything.
3. Beat eggs in another bowl and add milk there.
4. Put 2 cups of flour in the third bowl and add spices.
5. Put your chicken in flour with salt and pepper, after that cover the whole chicken with eggs and after that with the rest of flour.
6. Then put oil in the basket for Air Fryer and put chicken.
7. Cook it for 35 minutes at 360F, but after 20 minutes you should put it on another side.
8. Just enjoy with crispy, juicy and amazing chicken.

Nutrition:

- Calories: 372
- Fat: 15,3g
- Carbohydrates: 6,3g
- Protein: 36,2g

Chicken Sandwich

Sometimes, you can face the fact that chicken meal is too fry. However, here you will get only delicious and soft chicken.

Prep time: 40 minutes

Cooking time: 16 minutes

Servings: 2

Ingredients:

- ¼ teaspoon of red pepper for spicy sandwiches
- 7 dill pickle chips
- 4 toasted hamburger buns
- 1 oil mister
- 1/3 tablespoon of oil
- 1/2 teaspoon of celery
- 1/3teaspoon of garlic powder
- 1/3 teaspoon of black pepper
- 1 teaspoon of salt
- 1 teaspoon of paprika
- 1 teaspoon of powdered sugar
- 1 cup of flour
- 2 eggs
- ½ cup of milk
- ½ cup of dill pickle juice
- 2 chicken breasts (boneless and skinless)

Directions:

1. Cut chicken into the pieces.
2. Then put chicken in the bowl and rub with juice.
3. Leave if for 25-30 minutes.
4. Mix all eggs with milk in the bowl.
5. Put chicken in the bowl with milk.
6. Take another bowl and mix flour with all spices.
7. Then put chicken in this bowl.
8. Spray the backing basket for the Air Fryer with oil.
9. Put chicken in the Air Fryer and sprinkle with oil too.
10. Cook it for 6 minutes at 350F.
11. Then put chicken on another side, sprinkle with oil again and cook for 6 minutes more.
12. Increase the temperature at 360F and cook for 2 minutes for every side of chicken.
13. Serve hot and on buns. Also, it is possible to put 2 chips on every bun and if you like, you can add the ¼ teaspoon of mayonnaise on every bun. You will be surprised how delicious chicken can be.

Nutrition:

- Calories: 369
- Fat: 16,3g
- Carbohydrates: 7,3g
- Protein: 34,8g

Flavorsome Grilled Chicken

It has some unusual ingredients. Meat is juicy and crusty.

Prep time: 5 minutes

Cooking time: 1 hour

Servings: 4

Ingredients:

- 1 chicken (skinless)
- 1-2 tablespoons of oil or coconut oil
- 2 teaspoons of salt
- 3 teaspoons of celery
- 1 teaspoon of celery seeds
- 4 teaspoons of sugar
- ½ teaspoon of potato starch
- ½ teaspoon of garlic
- ½ teaspoon of onion powder
- 1 teaspoon of paprika
- ½ teaspoon of turmeric

Directions:

1. Wash and clean chicken.
2. Blend flavors with oil and rub chicken with this mixture.
3. Put chicken meat in the Air Fryer.
4. Cook chicken meat for 30 minutes at 360F.
5. Then place chicken on the other side.
6. Cook it for another 30 minutes.
7. After that, leave it for 10 minutes and serve hot.
8. Decorate with parsley and basil leaves.

Nutrition:

- Calories: 366
- Fat: 16,2g
- Carbohydrates: 6,4g
- Protein: 36,8g

Tasty Chicken Legs with Spices

Prepare this chicken. This meal is flavorsome and sharp.

Prep time: 2 hours

Cooking time: 20 minutes

Servings: 3

Ingredients:

- 1 cup of buttermilk
- 1 tablespoon of oil
- 1 tablespoon of paprika
- 1 teaspoon of cumin
- ½ teaspoon of poultry seasonings
- 1 teaspoon of onion powder
- 1 tablespoon of garlic powder
- 1 tablespoon of black pepper
- 2 cups of flour
- 3 chicken legs

Directions:

1. The legs of chicken should be placed in the container with buttermilk.
2. Then leave them for 2 hours. The poultry will have special note after that.
3. Blend the paprika, cumin, onion powder, black pepper and garlic powder.
4. Take the poultry out of fridge, place in flour.
5. Then cover chicken with buttermilk and two cups of flour.
6. After that set chicken legs in the fry basket in the Air Fryer.
7. Cook it for 20 minutes at 380F.
8. Turn the legs of chicken on the other side every 5-6 minutes.
9. After that serve hot and decorate with parsley or basil and mint leaves.

Nutrition:

- Calories: 385
- Fat: 16,5g
- Carbohydrates: 6,8g
- Protein: 36,4g

Healthy Chicken Tenders

Chicken will be tasty and juice if you use this recipe. However, you even cannot imagine how simply and quickly it can be prepared.

Prep time: 10 minutes

Cooking time: 10 minutes

Servings: 3

Ingredients:

- Salt
- Pepper
- 1,5 oz of bread crumbs
- 1/8 cup of flour
- 1 big egg
- 12 oz of chicken breasts

Directions:

1. Cut all unneeded fat from chicken.
2. Wash it under the cold water.
3. Rub chicken with salt and pepper.
4. Put chicken in flour and cover it with flour.
5. Take another bowl and beat eggs there.
6. After that put it in beaten eggs.
7. Then cover chicken with bread crumbs.
8. Sprinkle the Air Fryer basket with oil and put chicken there.
9. Cook for 10 minutes at 352F.
10. When chicken is ready, server it hot. You will appreciate this recipe and will prepare it very often.

Nutrition:

- Calories: 388
- Fat: 15,6g
- Carbohydrates: 6,8g
- Protein: 36,7g

Meat

Meat Burgers

If you wish to make the great and fast meat snacks, cook these fantastic burgers.

Prep time: 10 minutes
Cooking time: 10 minutes
Servings: 4

Ingredients:

- 1 pound of beef
- 1 teaspoon of dry parsley
- ½ teaspoon of oregano
- ½ teaspoon of salt
- ½ teaspoon of pepper
- ½ teaspoon of onion powder
- 1 tablespoon of sauce

Directions:

1. Put meat in the food processor and cut it.
2. Then put the beef in the bowl.
3. After that mix dry parsley, oregano, salt, pepper, onion powder and sauce in another bowl.
4. Then add all these spices to meat.
5. Mix everything well.
6. Make the small burgers.
7. Sprinkle the Air Fryer basket with oil.
8. Put the burgers in the Air Fryer.
9. Cook 10 minutes at 300F.
10. When the burgers are ready, serve them with buns, vegetables and sauces.

Nutrition:

- Calories: 148
- Fat: 4,6g
- Carbohydrates: 1,6g
- Protein: 24,2g

Balls with Bacon and Cheese

Do you need to prepare something interesting? The great mixture of cheese with bacon is very wonderful. Try to make and you will like it.

Prep time: 10 minutes

Cooking time: 10 minutes

Servings: 4

Ingredients:

- 1 cup of bread crumbs
- ¾ cup of flour
- 2 cups of milk
- 3 eggs
- 1/3 cup of cheese
- ¼ cup of bacon

Directions:

1. Beat three eggs in the bowl and mix them.
2. Chop cheese and bacon in the small pieces.
3. Add flour and mix everything well.
4. Make the balls from cheese and bacon with flour.
5. Take the ball, put it in milk and after that in bread crumbs.
6. All balls should be in bread crumbs and in milk.
7. After that preheat the Air Fryer to 350F.
8. Sprinkle the basket with oil.
9. Put the balls in the Air Fryer and cook for 5 minutes at 350F.
10. Then put them on another side and cook for 5 minutes more.
11. When the balls are ready, enjoy with them and add different sauces.

Nutrition:

- Calories: 165
- Fat: 5,6g
- Carbohydrates: 1,5g
- Protein: 24,3g

Beef Roll

It is very important to eat meat and because of this fact you can prepare the delicious, tasty and great meat. It will be great dinner for your family.

Prep time: 10 minutes

Cooking time: 14 minutes

Servings: 4

Ingredients:

- 1 teaspoon of pepper
- 1 teaspoon of salt
- ¾ cup of spinach
- 3 oz of red chili pepper
- 6 pieces of cheese
- 3 tablespoons of sauce
- 2 lb of beef steak

Directions:

1. Divide the steak in the separate pieces.
2. Rub the pieces of beef with salt and pepper.
3. Chop cheese, pepper and spinach into the small pieces.
4. Mix them together.
5. Then put the mixture on every piece of meat and roll them.
6. Sprinkle the Air Fryer basket with oil
7. Put the rolls in the Air Fryer.
8. Cook for 14 minutes at 380F.
9. Serve hot with sauce.
10. Enjoy with tasty rolls.

Nutrition:

- Calories: 156
- Fat: 5,8g
- Carbohydrates: 1,6g
- Protein: 24,6g

Beef Tenders

They are very easy for preparing and at the same moment they are very delicious. You can prepare the complete dinner for your family if you prepare this meal.

Prep time: 10 minutes

Cooking time: 15 minutes

Servings: 4

Ingredients:

- ½ cup of milk
- 1 tablespoon of oil
- ½ teaspoon of salt
- ½ teaspoon of oil
- 1 cup of bread crumbs
- ½ cup of flour
- 3 eggs
- 1 lb of meat

Directions:

1. Mix bread crumbs with oil.
2. Then take another bowl and mix eggs with milk.
3. Put flour in the third bowl.
4. Chop meat into middle pieces.
5. Take the piece of meat, put it in flour, then in beaten eggs and after that in bread crumbs.
6. Sprinkle the Air Fryer basket with oil.
7. Put the pieces of meet in the Air Fryer.
8. Cook them for 15 minutes at 380F.
9. Enjoy with this delicious meat.

Nutrition:

- Calories: 144
- Fat: 5,3g
- Carbohydrates: 1,2g
- Protein: 24,7g

20 Minutes Beef

This type of meat will be the beloved one, since it can be popular as for trip as for the typical dinner in 20 minutes only. You will like it as the beef is simple and pleasant. Just cook meat!

Prep time: 10 minutes

Cooking time: 20 minutes

Servings: 4

Ingredients:

- 1 egg
- ½ teaspoon of pepper
- ½ teaspoon of salt
- ½ teaspoon of cumin
- 1 tablespoon of tomato pasta
- 1 small onion
- 2 cloves of garlic
- 1 green pepper
- 1 tablespoon of oil
- 1 cup of flour

Directions:

1. Blend salt, pepper then cumin and garlic in the bowl.
2. Slice onion with pepper in small pieces.
3. Then mix them together.
4. Place oil and tomato pasta to the rest of products.
5. Mix them well.
6. Make cutlets from this mixture.
7. Sprinkle the Air Fryer basket with oil.
8. Then put cutlets in the Air Fryer.
9. Cook them for 15 minute at 300F and then put on other side and cook for 5 minutes.
10. Serve hot with sauce and vegetables.

Nutrition:

- Calories: 147
- Fat: 5,5g
- Carbohydrates: 1,3g
- Protein: 24,1g

Rib Eye Steak

If you wish to know how to prepare meat well and have what to show your guests, then this recipe is exactly for you. The guests will like it.

Prep time: 10 minutes

Cooking time: 21 minutes

Servings: 4

Ingredients:

- 1 teaspoon of oil
- ½ teaspoon of pepper
- ½ teaspoon of salt
- 1/3 teaspoon of dry herbs
- 2 lbs of steak rub

Directions:

1. Preheat the Air Fryer to 390F for 4 minutes.
2. Then mix salt, pepper, dry herbs and oil together.
3. After that rub meat with these spices.
4. Sprinkle the Air Fryer basket with oil.
5. Put meat in the Air Fryer.
6. Cook for 14 minutes at 390F.
7. Then put the steak on another side and cook for 7 minutes.
8. Serve hot.
9. Decorate with vegetables and do not forget about sauces. Your guests will be surprised, because it will be too tasty.

Nutrition:

- Calories: 138
- Fat: 4,9g
- Carbohydrates: 1,5g
- Protein: 23,8g

Garlic Meat

It is for fans of spicy food. This garlic meat is delicious and spicy. It is very rare and tasty. Do not miss your chance.

Prep time: 10 minutes

Cooking time: 12 minutes

Servings: 4

Ingredients:

- 1 lb of meat.
- ½ teaspoon of salt
- ½ teaspoon of pepper
- 1/3 teaspoon of red chili pepper
- ½ teaspoon of garlic
- ½ teaspoon of mustard
- 3 eggs
- 1 cup of flour

Directions:

1. Take the deeps bowl.
2. Mix salt, pepper then red chili pepper, garlic, mustard in this bowl.
3. Then cut meat in the pieces.
4. Rub every piece of meat with spices.
5. Beat three eggs in the bowl and put flour in another bowl.
6. Then put every piece of meat in flour and then in egg,
7. Sprinkle the frying basket with oil.
8. Put the steaks in the Air Fryer.
9. Cook then for 6 minutes at 380F and then 6 minutes on the other side.
10. Serve hot with vegetables.

Nutrition:

- Calories: 135
- Fat: 4,6g
- Carbohydrates: 1,7g
- Protein: 23,5g

Tater Tots with Bacon

This meal seems to be original, but you even cannot imagine how flavorsome it is. The result of the cooked bacon will exceed all your expectation.

Prep time: 10 minutes

Cooking time: 8 minutes

Servings: 4

Ingredients:

- ½ cup of cheese
- ½ teaspoon of salt
- ½ teaspoon of pepper
- 1/3 teaspoon of paprika
- 3 tablespoons of sour cream
- 4 onions
- Frozen tater tots

Directions:

1. Chop bacon in the pieces.
2. Mix salt, pepper then paprika in the bowl.
3. Chop onions and add to all ingredients.
4. Crush every tater tots and mix with cheese.
5. Add to the rest of products.
6. Mix everything well.
7. Preheat the Air Fryer to 380F.
8. Sprinkle the basket with oil.
9. Put the assortment in the Air Fryer and cook for 8 minutes at 380F.
10. Then add sour cream.
11. Serve hot!

Nutrition:

- Calories: 158
- Fat: 4,8g
- Carbohydrates: 1,4g
- Protein: 22,5g

Easy Steak

Prepare this scrumptious meat, the easy steak is your true choice. The family will be shocked once they try it.

Prep time: 10 minutes

Cooking time: 8 minutes

Servings: 4

Ingredients:

- 1 teaspoon of pepper
- ½ teaspoon of salt
- 2 cups of milk
- 2 tablespoons of flour
- 1 teaspoon of garlic powder
- 1 teaspoon of onion powder
- 1 cup of flour
- 1 cup of bread crumbs
- 3 beaten eggs
- 6 oz of steak

Directions:

1. Take the bowl and mix bread crumbs with salt, pepper, onion powder and garlic powder.
2. Rub the pieces of steal in this mixture.
3. Then beat eggs in the bowl.
4. Put the steak in flour, then in beaten eggs and after that in bread crumbs.
5. Sprinkle the Air Fryer with oil.
6. Put meat in Air Fryer.
7. Cook for 12 minutes at 350F.
8. When it is ready, serve with sauce and enjoy.

Nutrition:

- Calories: 138
- Fat: 4,3g
- Carbohydrates: 1,2g
- Protein: 22,7g

Wonderful Potatoes with Scrumptious Pork

This pork meal is for dinner. It is delightful, uncomplicated and the potato is flavorsome.

Prep time: 10 minutes

Cooking time: 14 minutes

Servings: 4

Ingredients:

- 1 tablespoon of balsamic glaze
- 1 teaspoon of salt
- 1 teaspoon of parsley
- 1 teaspoon of pepper
- ½ teaspoon of onion powder
- ½ teaspoon of garlic powder
- ½ teaspoon of red chili pepper
- 1 red potatoes
- 2 lb of meat

Directions:

1. Wash and clean potatoes.
2. Slice the products in the pieces.
3. Rub them with salt, pepper, parsley then onion powder, garlic powder and red chili pepper.
4. Lay meat in the Air Fryer and cook for 7 minutes at 380F.
5. Then put meat on the dish.
6. Cook vegetables in the Air Fryer for 3 minutes and then 4 minutes on the other side.
7. Place all components on saucer, decorate with the root vegetables and pasta or basil leaves with parsley.

Nutrition:

- Calories: 147
- Fat: 4,6g
- Carbohydrates: 1,4g
- Protein: 22,76g

Hot Dogs

These hot dogs are the special ones. Attempt to create these hot dogs and you can be satisfied with the result. Your effort will be valued by the children.

Prep time: 10 minutes

Cooking time: 3 minutes

Servings: 4

Ingredients:

- 1 tablespoon of ketchup
- 8 candies eyes
- 4 hot dogs
- ½ teaspoon of pepper
- ½ teaspoon of salt
- 2 tablespoons of oil
- 1 puff pastry sheet

Directions:

1. Rub sausages with salt and pepper.
2. Place the puff pastry sheet on the table.
3. Divide it in 5 strings.
4. Cover sausages with these strings.
5. Sprinkle the Air Fryer with oil.
6. Set the hot dogs in the Air Fryer for 3 minutes at 300F.
7. Then add the candies eyes.
8. Prepare hot with sauces. You can choose different ones.
9. Your kids will like them.
10. Cook and you will be delighted with it.

Nutrition:

- Calories: 119
- Fat: 4,1g
- Carbohydrates: 1,2g
- Protein: 22,1g

Turkey with Spices

Do you need to cook soft and tasty meat? Cook this turkey and you will use too little time preparing meat.

Prep time: 15 minutes

Cooking time: 40 minutes

Servings: 4

Ingredients:

- 2 tablespoons of olive oil
- 1 teaspoon of salt
- 1 teaspoon of pepper
- 8 pounds of turkey.

Directions:

1. Take the bowl and mix salt, pepper and oil in it.
2. Rub turkey with these seasonings and oil.
3. Chop turkey in the pieces. It will be easier for you to cook if they will not be too big.
4. Then sprinkle the frying basket with oil.
5. Put the pieces of turkey in the Air Fryer.
6. Cook them for 20 minutes at 280F.
7. Then put them on other side and cook for 20 minutes again.
8. Put turkey in the plate.
9. Decorate with basil leaves, mint and do not forget about sauces.
10. It will be your favorite and delicious meal.

Nutrition:

- Calories: 138
- Fat: 4,2g
- Carbohydrates: 1,4g
- Protein: 22,5g

Meat with Herbs

It is enjoyable and peppery meal. It can be roasted for breakfast, for dinner and for lunch. Just attempt and you will definitely like this meat.

Prep time: 10 minutes

Cooking time: 20 minutes

Servings: 4

Ingredients:

- ½ teaspoon of salt
- ½ teaspoon of pepper
- ½ teaspoon of onion powder
- ½ teaspoon of garlic
- 1/3 teaspoon of cumin
- 2 lb of meat
- 7 oz of cheese

Directions:

1. Take the bowl and mix salt, pepper, onion powder, garlic and cumin in it.
2. Cut meat into the pieces and rub them with spices.
3. After that preheat the Air Fryer to 380F.
4. Sprinkle the basket with oil.
5. Put the pieces of meat in the Air Fryer.
6. Cook for 10 minutes at 380F and then shake well and cook for 10 minutes more.
7. Put on the plate and serve with potatoes and sauces.
8. Meat should be hot, it is the most delicious at this moment.

Nutrition:

- Calories: 146
- Fat: 4,7g
- Carbohydrates: 1,3g
- Protein: 22,5g

Meet Cheese

This meat with cheese is uncomplicated in preparing: just 7 minutes. Unassuming, easy and scrumptious meal.

Prep time: 10 minutes

Cooking time: 7 minutes

Servings: 4

Ingredients:

- ½ teaspoon of salt
- ½ teaspoon of garlic powder
- ½ teaspoon of pepper
- ½ teaspoon of oregano
- 1/3 teaspoon of cumin
- ½ teaspoon of onion powder
- 8 pieces of cheese
- 2 tablespoons of oil
- 4 tablespoons of coleslaw
- 4 pieces of bread
- 1 lb of meat

Directions:

1. Blend oil with bread.
2. Place bread on the dish.
3. Mix all flavors together in the bowl.
4. Rub meat with them.
5. Put the piece of meat on bread.
6. Then put the coleslaw on it.
7. After that sprinkle the Air Fryer basket with oil.
8. Preheat it to 380F.
9. Put the snacks in the Air Fryer and cook for 5 minutes at 380F.
10. Then add pieces of cheese and cook 2 minutes at 300F.
11. Cook hot and enjoy with the tasty meal.

Nutrition:

- Calories: 165
- Fat: 5,7g
- Carbohydrates: 1,2g
- Protein: 23,5g

Meatlof

If you wish to prepare something interesting and delicious at the same time, you should prepare this meal.

Prep time: 10 minutes

Cooking time: 20 minutes

Servings: 4

Ingredients:

- 2 mushrooms
- ½ teaspoon of salt
- ½ teaspoon of oil
- ½ teaspoon of pepper
- 3 oz of salami
- 1 onion
- 1 beaten egg
- 3 tablespoons of bread crumbs

Directions:

1. Cut the salami into pieces and put in the bowl.
2. Then add salt, oil, pepper there.
3. Chop onion and add to the ingredient.
4. Mix everything well.
5. Beat egg and add chopped mushrooms in the bowl.
6. Mix all ingredients.
7. After that, sprinkle the Air Fryer basket with oil.
8. Then make the cutlet from the mixture, put them in bread crumbs and after that in the Air Fryer.
9. Cook for 10 minutes at 380F and them 10 minutes more at 200F.
10. Serve hot with potatoes and sauces.

Nutrition:

- Calories: 136
- Fat: 5,4g
- Carbohydrates: 1,3g
- Protein: 23,4g

Vegetable Meals

Flavorsome Cooked Avocado

It is very simple, easy to prepare and most importantly - an incredibly delicious meal.

Prep time:10 minutes

Cooking time:10 minutes

Servings:4

Ingredients:

- ½ teaspoon of salt
- ½ teaspoon of pepper
- 1 big avocado
- 2 oz of white beans
- ½ cup of bread crumbs

Directions:

1. First of all, take the deep bowl.
2. Mix pepper, salt and bread crumbs.
3. Then place white beans in another bowl.
4. Chop avocado in small pieces.
5. Take the piece of avocado and put it in beans and after that in bread crumbs.
6. Preheat the Air Fryer to 390F.
7. Sprinkle the frying basket with oil.
8. After that put avocado pieces in the Air Fryer.
9. Cook for 5 minutes.
10. After that, shake avocado pieces well and cook for 5 minutes.
11. Serve hot and also you can take your favorite sauce.
12. Enjoy!

Nutrition:

- Calories: 212
- Fat: 20g
- Carbohydrates: 6g
- Protein: 2g

Apple Chips

The snacks, which are very easy for preparing. Also, these chips are very delicious.

Prep time:10 minutes

Cooking time:15 minutes

Servings:2

Ingredients:

- ¼ teaspoon of salt
- 1 tablespoon of sugar
- ½ teaspoon of cinnamon
- 1 big apple

Directions:

1. Firstly, preheat the Air Fryer to 380F.
2. Wash and clean the apple.
3. Chop the apple into the pieces.
4. Take small bowl and mix salt, sugar and cinnamon well.
5. Sprinkle the fry basket with oil or special spray.
6. Put the pieces of apple in the Air Fryer.
7. Cover the apple with salt, sugar and cinnamon.
8. Cook 7-8 minutes at 380F in Air Fryer.
9. Then put the pieces of the apple on another side and cook for 7-8 minutes again.
10. Then put them on the plate and wait 3-4 minutes till they are warm.
11. Eat them with orange juice.

Nutrition:

- Calories: 253
- Fat: 0,2g
- Carbohydrates: 59g
- Protein: 2,2g

Crispy Tofu

The incredibly delicious meal. All your friends will like it.

Prep time:35 minutes

Cooking time:20 minutes

Servings:2

Ingredients:

- 1 tablespoon of potato starch
- 2 teaspoons of sesame oil
- ¼ teaspoon of salt
- ¼ teaspoon of pepper
- 1 teaspoon of vinegar
- 2 tablespoons of soy sauce
- 1 block of tofu

Directions:

1. Take the bowl.
2. Chop tofu into small pieces.
3. Put them in the bowl.
4. After that, add sesame oil, teaspoon of vinegar and mix everything well.
5. Then add soy sauce and mix everything again.
6. Also, put pepper and salt to the mixture.
7. Leave everything for 15-30 minutes.
8. After that, put the potato starch to tofu and mix together.
9. Put the frying basket.
10. Add oil there.
11. Then put the pieces of tofu in the Air Fryer.
12. Cook for 10 minutes at 370F.
13. Then shake well and cook for 10 minutes more.

Nutrition:

- Calories: 76
- Fat: 4,78g
- Carbohydrates: 1,88g
- Protein: 8,8g

Small Breakfast Burritos

If you wish to prepare something special for breakfast, then you made the right choice. These burritos are tasty and delicious.

Prep time:15 minutes

Cooking time:25 minutes

Servings:2

Ingredients:

- Pinch of spinach
- 6-8 fresh asparagus
- 1 small broccoli
- 8 strips of red pepper
- 1/3 cup of sweet potato
- 2 tofu scrambles
- 4 pieces of rice paper
- 1-2 tablespoons of water
- 1-2 tablespoons of liquid smoke
- 2-3 tablespoons of tamari
- 2 tablespoons of cashew butter

Directions:

1. Preheat the Air Fryer to 355F.
2. Put rice paper in Air Fryer.
3. Take small bowl.
4. Mix together liquid smoke, water, tamari and cashew butter.
5. Then chop asparagus, spinach, red pepper, sweet potato, tofu in the bowl and mix with the rest ingredients, which are in the bowl.
6. Put everything on rice paper and cook in the Air Fryer.
7. The burritos should be cooked at 350F for 10 minutes.
8. Then put the burritos on another side and cook for 15 minutes.
9. Serve warm and enjoy with the delicious snacks.

Nutrition:

- Calories: 132
- Fat: 8,5g
- Carbohydrates: 1,4g
- Protein: 8,3g

Ranch Kale Chips

These chips are very easy for preparing. The result will be great.

Prep time:5 minutes

Cooking time:5 minutes

Servings:4

Ingredients:

- ¼ teaspoon of salt
- ¼ teaspoon of oregano
- 1/3 teaspoon of garlic
- ½ teaspoon of turmeric
- 1 tablespoon of yeast flakes
- 2 teaspoon of seasonings
- 4 cups of kale
- 2 tablespoons of oil

Directions:

1. Preheat the Air Fryer to 370F.
2. Cut the kale into the pieces.
3. Then mix the kale with salt, oregano, turmeric, garlic powder, oil and yeast flakes in the bowl.
4. Put the kale in the Air Fryer.
5. Then cook it for 4-5 minutes in the Air Fryer.
6. It should have dark green color.
7. After that enjoy with the healthy and delicious kale.
8. It is possible to prepare with potatoes or decorate with the different vegetables.

Nutrition:

- Calories: 27
- Fat: 1g
- Carbohydrates: 4,7g
- Protein: 1,8g

Taco Crisp Wraps

It is tasty meal, which can be prepared as snack for guests. Easy in cooking and delicious.

Prep time:10 minutes

Cooking time:18 minutes

Servings:4

Ingredients:

- 4 tablespoons of vegan cheese
- Mixed greens
- Tortilla chips
- 1/3 cup of mango salsa
- 4 pieces of fish fillet
- 2 cobs of grilled corn
- 1 red pepper
- 1 yellow onion
- 4 burrito tortillas

Directions:

1. Preheat the Air Fryer to 390F.
2. Cut cheese and fish fillet in the pieces.
3. Then add the corn to it and mix everything.
4. Then cut the red pepper and onion.
5. Mix everything well with chips.
6. Then sprinkle the Air Fryer basket with oil.
7. Take the burrito tortillas and divide the mixture in the same parts.
8. Put this mixture on tortillas.
9. Then roll them and put in the Air Fryer.
10. Cook for 12 minutes at 390F.
11. Then shake everything and cook for 6 minutes more.
12. Enjoy with delicious burritos and you can take your favorite sauce.

Nutrition:

- Calories: 238
- Fat: 13,7g
- Carbohydrates: 9,6g
- Protein: 3,7g

Chonut Holes

This meal is tasty and delicious. It is very easy and simple in preparing. Your friends will like it.

Prep time:1h 15 minutes

Cooking time:11 minutes

Servings:4

Ingredients:

- 2 tablespoons of sugar
- 2 teaspoons of cinnamon
- ¼ cup of almond milk
- 2 tablespoons of aquafaba
- ½ teaspoon of salt
- 1 teaspoon of backing powder
- 1 cup of flour
- ¼ cup of sugar
- 1 red pepper

Directions:

1. Take the deep bowl.
2. Mix salt, aquafaba, sugar and ¼ cup of sugar together in the bowl.
3. Put the mixture aside for 1 hour.
4. Cut the red pepper in small pieces.
5. Mix the cinnamon and 2 tablespoons of sugar in the separate bowl.
6. After that, mix everything together with pepper.
7. Divide the mixture in 12 pieces and create the rolls.
8. Put the rolls in the Air Fryer.
9. Cook for 6 minutes at 370F.
10. Then shake them and cook 5 minutes more.

Nutrition:

- Calories: 197
- Fat: 13,2g
- Carbohydrates: 6,6g
- Protein: 2,7g

Tofu Scramble

This recipe will be the new one in your collection. Tasty and delicious. Just try!

Prep time: 5 minutes

Cooking time:30 minutes

Servings:4

Ingredients:

- 2 cups of broccoli
- 1 tablespoon of oil
- 2 cups of red potatoes
- ½ cup of onion
- ½ teaspoon of onion powder
- ½ teaspoon of garlic powder
- 1 teaspoon of turmeric
- 1 tablespoon of olive oil
- 2 tablespoons of sauce
- 1 block of tofu

Directions:

1. Chop broccoli, potatoes, onion.
2. Put them in the bowl.
3. Then add tofu, oil, onion powder, garlic powder and mix together.
4. Leave the mixture aside.
5. After that, put the mixture in the Air Fryer.
6. Do not forget to sprinkle the basket with oil.
7. Cook for 7-8 minutes at 350F.
8. Then put the mixture on another side and cook for 15 minutes at 370F.
9. Serve it hot.
10. You can decorate with the different vegetables or basil leaves.

Nutrition:

- Calories: 225
- Fat: 16,2g
- Carbohydrates: 7,6g
- Protein: 4,7g

Flavorsome Cauliflower

A fine dietary dinner or a side dish to the main course! It will be the great choice.

Prep time: 5 minutes

Cooking time:20-22 minutes

Servings:4

Ingredients:

- 4 cups of cauliflower
- 2 carrots
- 1 cup of bread crumbs
- 1 teaspoon of salt
- ¼ cup of vegan sauce
- ¼ cup of buffalo sauce
- ¼ teaspoon of pepper

Directions:

1. Mix the vegan sauce with the buffalo sauce.
2. Put every cauliflower in the mixture.
3. Then add salt, pepper and mix everything.
4. After that, chop carrots.
5. Put every cauliflower in bread crumbs and after that put in the Air Fryer.
6. Add carrot and cook for 14-17 minutes at 370F.
7. After that, shake everything well and cook for another 5 minutes.
8. When everything is ready, enjoy with the healthy and tasty food.

Nutrition:

- Calories: 139
- Fat: 15,2g
- Carbohydrates: 4,3g
- Protein: 4,2g

Brussels sprouts

It is really healthy meal. Also, you will see, that it is very delicious and children will like it.

Prep time: 5 minutes

Cooking time:10 minutes

Servings:4

Ingredients:

- ¼ teaspoon of salt
- ¼ teaspoon of pepper
- 1 tablespoon of oil
- ¼ teaspoon of paprika
- 1/3 teaspoon of oregano
- ½ teaspoon of garlic
- 1/3 teaspoon of ginger
- 1/3 teaspoon of cumin
- 2 cups of brussels sprouts

Directions:

1. Preheat the Air Fryer to 380F.
2. Put salt, pepper, oil, oregano and paprika in the bowl.
3. Mix everything well.
4. After that add garlic, cumin and ginger.
5. Mix these ingredients with the rest in the bowl.
6. You can cut brussels sprouts in 2 pieces.
7. Then rub every piece with spices.
8. Sprinkle the frying basket with oil.
9. Put brussels sprouts in the basket.
10. Cook for 8 minutes and then shake them.
11. After that, cook for another 8 minutes.
12. Serve hot and enjoy with healthy and crispy brussels sprouts with your family.

Nutrition:

- Calories: 86
- Fat: 6,2g
- Carbohydrates: 4,1g
- Protein: 2,2g

Piquant Chickpeas with Juice of Lemon

It is appetizing and vigorous meal. This meal is delectable and brilliant.

Prep time: 5 minutes

Cooking time:20 minutes

Servings:4

Ingredients:

- 2,5 tablespoons of juice of the fresh lemon
- ½ teaspoon of salt
- ½ teaspoon of pepper
- ½ teaspoon of paprika
- ½ teaspoon of cumin
- ½ teaspoon of oregano
- 2 tablespoons of oil
- 15 oz of chickpeas

Directions:

1. Take the small bowl.
2. After that blend the chickpeas with the fat in the container. When you are purchasing the chickpeas, you should not choose very big size.
3. Cook up the Air Fryer to 390F.
4. Then cook the chickpeas up to 15 minutes.
5. Later add rest of oil, oregano, cumin, paprika, then pepper, juice of lemon and salt.
6. Blend the foods in the bowl.
7. Place the chickpeas in the Air Fryer and cook for 5 minutes.
8. Serve with salad or vegetables.
9. Enjoy with the meal.

Nutrition:

- Calories: 92
- Fat: 3,3g
- Carbohydrates: 3,2g
- Protein: 2,1g

Tofu

You can have the totally original and unusual meal, which will become your favored. There are countless minerals and different vitamins in this tofu meal.

Prep time: 40 minutes

Cooking time:30 minutes

Servings:4

Ingredients:

- 1 teaspoon of salt
- 1 cup of bread crumbs
- ½ cup of vegan mayo
- 1 teaspoon of ginger
- ½ teaspoon of garlic
- 1 teaspoon of vinegar
- ¼ cup of soy sauce
- 1 tablespoon of sesame oil
- 1 block of tofu

Directions:

1. Make 8 cutlets of tofu.
2. Make marinade.
3. Mix sesame oil, soy sauce, garlic with vinegar and ginger together in the container.
4. Set cutlets on the plate and cover with marinade.
5. Leave the products for 30 minutes.
6. Place the vegan mayo in the bowl.
7. Mix salt and bread crumbs in another salt.
8. Put cheese in the mayo and then in bread crumbs.
9. Cook then for 20 minutes at 350F, then shake and cook for 10 minutes more.
10. You will get tasty cutlets, which can be good as for dinner as for breakfast.

Nutrition:

- Calories: 103
- Fat: 8,4g
- Carbohydrates: 5,1g
- Protein: 3,1g

Appetizing Potato

It is yummy meal and it is uncomplicated for preparing. Try and see.

Prep time: 10 minutes

Cooking time: 35-40 minutes

Servings: 4

Ingredients:

- 1 teaspoon of salt
- 1 teaspoon of pepper
- 1 tablespoon of chives
- 1 tablespoon of Kalamata olives
- 1 piece of bacon
- 1 cup of vegan cream cheese
- 1/8 teaspoon of salt
- ¼ teaspoon of onion powder
- 1 teaspoon of oil
- 1 medium Russet potato

Directions:

1. Clean and wash potatoes
2. Then rub it with oil, onion powder and 1/8 teaspoon of salt.
3. Preheat the Air Fryer to 390F.
4. Put potatoes in the Air Fryer basket.
5. Cook potatoes for 35 minutes.
6. The potato should have the wonderful coffee color. Then place it on the other side and cook further.
7. Then add the rest of foodstuffs and cook for 5 minutes more.
8. Decorate with pasta and basil leaves.

Nutrition:

- Calories: 272
- Fat: 20.1g
- Carbohydrates: 37.1g
- Protein: 33.2g

Crispy Fried Pickles

You will get the most delicious and tasty meal. Enjoy with crispy skin of meal.

Prep time: 10 minutes

Cooking time: 35-40 minutes

Servings: 4

Ingredients:

- ½ cup of vegan sauce
- 2 teaspoons of oil
- ¼ teaspoon of cayenne pepper
- ½ teaspoon of paprika
- 6 tablespoons of bread crumbs
- 2 tablespoons of corn starch
- 2-3 tablespoons of water
- ¼ teaspoon of salt
- 3 tablespoons of dark beer
- 1/8 teaspoon of baking powder
- ¼ cup of flour
- 14 pickle slices

Directions:

1. Wash and clean the pickle slices, dry and leave aside.
2. Mix the baking powder, beer, salt and water together.
3. Then take two plates.
4. Put the corn starch on the first plate.
5. Mix the paprika, salt, cayenne pepper and bread crumbs on the second plate.
6. Put every piece in the corn starch and after that in beer batter.
7. After that, put everything in bread crumbs.
8. Preheat the Air Fryer to 380F.
9. Cook in the Air Fryer at 8 minutes and then shake well.
10. Cook for another 8 minutes.
11. All pieces should have brown color.
12. When everything is ready, serve with the different sauces to taste.

Nutrition:

- Calories: 225
- Fat: 16.1g
- Carbohydrates: 33.1g
- Protein: 31.2g

Vegetable Fries

This meal will be great if you wish to prepare something easy and healthy. Everyone will like it, even children.

Prep time: 5 minutes

Cooking time: 15-20 minutes

Servings: 4

Ingredients:

- ¼ teaspoon of salt
- ¼ teaspoon of pepper
- 1 teaspoon of basil
- 1 tablespoon of mix spices
- 1 tablespoon if thyme
- 2 tablespoons of olive oil
- 6 oz of carrot
- 8 oz of courgette
- 8 oz of sweet potatoes
- ¼ teaspoon of paprika
- ¼ teaspoon of oregano
- 1/3 teaspoon of chili

Directions:

1. Clean and wash carrot and sweet potatoes.
2. Then clean the courgetti and chop them in the small pieces.
3. After that, chop the sweet potatoes and carrot into the slices.
4. Mix everything together.
5. After that, add salt, pepper, basil, mix spices, thyme, chili, pepper and oregano.
6. Sprinkle the frying basket with oil and put carrot, sweet potatoes and courgetti in the Air Fryer.
7. Cook them for 18 minutes.
8. Shake them after 5 and 12 minutes of the cooking.
9. Serve hot and enjoy with the healthy vegetables.

Nutrition:

- Calories: 89
- Fat: 9.1g
- Carbohydrates: 25.1g
- Protein: 23.2g

Desserts

Apricot and Blackberry Crumble

Minimum of products and the maximum of satisfaction. It is something incredible and tasty.

Prep time: 10 minutes

Cooking time: 20 minutes

Servings: 4

Ingredients:

- 2 oz of cold butter
- 3 oz of flour
- 1 tablespoon of lemon juice
- 3 oz of fresh blackberries
- 2 oz of sugar
- 7 oz of apricots
- 1 pinch of salt

Directions:

1. Preheat the Air Fryer to 390F.
2. Chop apricots into the pieces and mix them with sugar and lemon juice in the bowl.
3. Grease the cake tin and put fruits on it.
4. Mix flour with salt and butter and add water.
5. Put the mixture on the fruits.
6. After that put it in the frying basket and cook in the Air Fryer for 20 minutes.
7. Dessert should have warm golden color.
8. Serve with ice cream or jam.
9. Eat and enjoy with it.

Nutrition:

- Calories: 480
- Fat: 30,51g
- Carbohydrates: 46,11g
- Protein: 6,2g

Strawberry Cupcakes

Recipe for a very delicate and light dessert. You will see, that it is very delicious.

Prep time: 10 minutes

Cooking time: 8 minutes

Servings: 4

Ingredients:

- ¼ cup of fresh strawberries
- 1tablespoon of whipped cream
- ½ tablespoon of pink food coloring
- 4 oz of butter
- ½ tablespoon of vanilla essence
- 3 oz of flour
- 2 eggs
- 3 oz of sugar
- 3 oz of butter

Directions:

1. Preheat the Air Fryer to 380F.
2. Mix butter and sugar in the bowl.
3. Then add vanilla essence and beat eggs one by one.
4. Then add some flour.
5. Create buns and cook them for 8 minutes in the Air Fryer at 350F.
6. Mix sugar, food coloring, whipped cream and strawberries well.
7. When the cupcakes are ready, decorate them with the mixture of cream.
8. Serve hot with tea or coffee.
9. You can be sure, that everyone will like them.

Nutrition:

- Calories: 630
- Fat: 45,5g
- Carbohydrates: 44,11g
- Protein: 6,1g

Tasty Profiteroles with Chocolate

It is tasty cake and dessert will be cooked easily. The products are affordable and inexpensive.

Prep time: 10 minutes

Cooking time: 10 minutes

Servings: 4

Ingredients:

- 2 oz of butter
- 2 tablespoons of whipped cream
- 3 oz of milk chocolate
- 2 tablespoons of sugar
- 2 tablespoons of vanilla essence
- 2 cups of water
- 6 eggs
- 6 oz of flour
- 3 oz of butter

Directions:

1. Beat eggs in the bowl.
2. Then mix them with flour.
3. After that, leave them aside.
4. Then create cream for dessert.
5. Mix vanilla essence, whipped cream and sugar together.
6. Put the dough in the Air Fryer and cook 10 minutes at 380F.
7. Melt chocolate and put it aside.
8. Decorate the ready cakes with cream and add chocolate above.
9. Enjoy with the tasty dessert.

Nutrition:

- Calories: 589
- Fat: 42,5g
- Carbohydrates: 43,11g
- Protein: 6,5g

British Sponge

Very tasty and easy dessert for tea. There is nothing easier to prepare. Try and you will like it forever.

Prep time: 15 minutes

Cooking time: 25 minutes

Servings: 8

Ingredients:

- 1 tablespoon of whipped cream
- 3 oz of icing sugar
- 2 oz of butter
- 2 tablespoons of strawberry jam
- 2 eggs
- 3 oz of sugar
- 3 oz of butter
- 3 oz of flour

Directions:

1. Preheat the Air Fryer to 350F.
2. Beat butter and sugar well.
3. Then beat eggs and add some flour.
4. Put the mixture in the Air Fryer and cook for 15 minutes at 350F and then 10 minutes at 340F.
5. Mix butter with the ice sugar and you will get creamy mixture.
6. Then cover dessert with jam and later with cream.
7. Enjoy with the delicious dessert and juice. All your friends will like this dessert because it is tasty.

Nutrition:

- Calories: 598
- Fat: 44,7g
- Carbohydrates: 41,4g
- Protein: 6,3g

Orange Chocolate Fondant

Very easy, tasty and delicious recipe, which you can prepare it always. You should not spend too much time for it.

Prep time: 15 minutes

Cooking time: 22 minutes

Servings: 4

Ingredients:

- 2 eggs
- 1 orange
- 3 oz of butter
- 3 oz of dark chocolate
- 4 teaspoons of sugar
- 2 tablespoons of flour

Directions:

1. Preheat the Air Fryer to 390F.
2. Melt chocolate and butter.
3. Then whist sugar and eggs together.
4. After that, add chopped orange with egg and sugar in chocolate.
5. Then add flour and mix all ingredients.
6. Divide the mixture in 12 pieces and cook them for 12 minutes in the Air Fryer at 390F.
7. Put the cakes on the plate and wait for 10 minutes. They should not be hot.
8. Serve with cold and scrumptious ice cream and juice.
9. Cook and enjoy with this dessert.

Nutrition:

- Calories: 585
- Fat: 45,8g
- Carbohydrates: 42,4g
- Protein: 7,3g

Heart-Shaped Churros

It will be your favorite cake. This cake is very easy for cooking. Also, you will find it delicious.

Prep time: 16-21 minutes

Cooking time: 6 minutes

Servings: 4

Ingredients:

- 1 cup of flour
- ½ cup of sugar
- ¼ teaspoon of salt
- ½ cup of better
- ½ cup of water

Directions:

1. Mix water with butter in the pan and melt it.
2. After that, put the mixture of butter and water in the bowl.
3. Then add flour and mix everything well. Create the dough.
4. Leave for 10-15 minutes.
5. After that, add egg, sugar and salt.
6. Mix everything well.
7. Then create the churros in the shape of heart.
8. Put them in the Air Fryer at 380F.
9. Cook them for 6 minutes.
10. Put the ready cakes on the plate.
11. You can combine them with different jams and enjoy with your friends.

Nutrition:

- Calories: 278
- Fat: 19,5g
- Carbohydrates: 29,4g
- Protein: 6,3g

Peanut Banana Butter Dessert Bites

Here you can find a lot of ingredients and because of it, this dessert is incredibly tasty. Just cook and you will see, that your family will eat this cake in 1 minute.

Prep time: 5-10 minutes

Cooking time: 6 minutes

Servings: 4

Ingredients:

- ¼ teaspoon of cinnamon
- M and M'S
- 3 tablespoons of raisins
- 3 tablespoons of chocolate chips
- 1 oil mister
- 2 teaspoons of oil
- ½ cup of peanut butter
- 1 large banana
- 1 teaspoon of lemon juice

Directions:

1. Chop the banana in the pieces and put in water with teaspoon of lemon.
2. After that, mix the cinnamon, M and M's, raisins, chocolate chips, peanut butter in the bowl.
3. Add the banana to this mixture.
4. After that preheat the Air Fryer to 380F.
5. Divide the mixture in the pieces and create balls.
6. Sprinkle the Air Fryer basket with 2 teaspoons of oil.
7. Put the sweet balls in the Air Fryer.
8. Cook for 6 minutes.
9. When they are ready, add ice cream or it is possible to cover them with the cinnamon.
10. Do not forget about juice!

Nutrition:

- Calories: 235
- Fat: 12,6g
- Carbohydrates: 25,3g
- Protein: 8,3g

Sweet Potato Pie

Can you imagine the pie from potato? No? Just try and you will see, that it is the best pie you have ever tasted.

Prep time: 30 minutes

Cooking time: 60 minutes

Servings: 4

Ingredients:

- 1 cup of whipped cream
- 1/8 teaspoon of ground nutmeg
- ½ teaspoon of cinnamon
- ½ teaspoon of salt
- ¾ teaspoon of vanilla
- 1 teaspoon of oil
- 1 tablespoon of sugar
- 2 tablespoons of maple syrup
- ¼ cup of cream
- 1 prepared pie
- 1 teaspoon of oil
- 1 6oz of sweet potatoes

Directions:

1. Wash and clean the sweet potatoes.
2. Cut them into pieces.
3. Sprinkle the Air Fryer basket with oil and cook potatoes for 30 minutes at 380F.
4. Do not forget to shake it after 15 minutes of cooking.
5. Mix potatoes with the ground nutmeg, cinnamon, salt, vanilla, sugar, maple syrup and cream.
6. Then put this mixture on the pie.
7. Cook in the Air Fryer at 320F for 30 minutes.
8. The pie should have golden brown color.
9. When it is ready, leave for 20 minutes.
10. Decorate with cream and enjoy!

Nutrition:

- Calories: 467
- Fat: 19,2g
- Carbohydrates: 22,3g
- Protein: 7,3g

Cinnamon Sugar Donuts

This is one of the most delicious, tender, high and porous donut recipe you have ever made. Rich chocolate taste, simplicity of cooking and a set of products that every housewife has at home!

Prep time: 5 minutes

Cooking time: 4 minutes

Servings: 2

Ingredients:

- 2-3 tablespoons of water
- ¼ cup of confectionary sugar
- 4 teaspoons of cinnamon
- 8 buttermilk biscuits
- 1 piece of chocolate

Directions:

1. Put all biscuits on the plate.
2. Mix sugar with the cinnamon and water
3. Put this mixture on every biscuit.
4. Melt chocolate.
5. Put the biscuits in the Air Fryer and cook for 4 minutes at 300F.
6. Then put on the plate and pour with chocolate.
7. Serve warm and it is possible to prepare milk of coffee with these donuts.
8. Eat them and enjoy with the result.

Nutrition:

- Calories: 414
- Fat: 14,2g
- Carbohydrates: 22,5g
- Protein: 7,2g

Peanut Butter Banana Smoothie

There is nothing easier than to cook this peanut banana smoothie. You will like it. Cook and taste and enjoy with it.

Prep time: 5 minutes

Cooking time: 2-3 minutes

Servings: 2

Ingredients:

- 2 tablespoons of ground flaxseed
- ¾ cup of vanilla yogurt (or you can choose any other you like)
- ½ cup of almond milk
- 1 banana
- 1 tablespoon of honey
- 1 tablespoon of peanut butter

Directions:

1. Chop the banana in the pieces.
2. Mix with yogurt, milk, honey, peanut butter and ground flaxseed well in the bowl.
3. Add the banana to the mixture of ingredients.
4. Preheat the Air Fryer to 380F.
5. Put the ingredients in the Air Fryer and then cook for 2-3 minutes.
6. Serve hot with fruits.
7. Also, you can put some ice and drink cold. It will be also delicious.

Nutrition:

- Calories: 56
- Fat: 1,2g
- Carbohydrates: 5,9g
- Protein: 5,2g

New York Cheesecake

This is something extraordinary! It is an indispensable dessert for your guests. It does not need a lot of expenses, has a very airy texture and most importantly – it does not take too much of your time.

Prep time: 20 minutes

Cooking time: 30 minutes

Servings: 4

Ingredients:

- 1 tablespoon of vanilla
- 3 eggs
- 2 cups of sugar
- 2 lbs of soft cheese
- 2 oz of melted butter
- 3 oz of butter
- 3 oz of brown sugar
- 7 oz of flour

Directions:

1. Mix flour with sugar well in the bowl.
2. Create the biscuit forms and put in the Air Fryer.
3. Cook for 15 minutes at 370F.
4. When the biscuits are ready, chop them in the small pieces.
5. Then add cheese, butter and sugar and whish everything.
6. You should get creamy mixture.
7. Beat eggs in the bowl and add vanilla.
8. After that, mix everything.
9. Put the mixture in the Air Fryer.
10. Cook for 30 minutes at 380F.
11. Serve hot with juice, tea or coffee.

Nutrition:

- Calories: 389
- Fat: 10,7g
- Carbohydrates: 9,9g
- Protein: 7,2g

Lime Cheesecake

Do you like lime? Prepare this lime cheesecake! The result will be great.

Prep time: 25 minutes

Cooking time: 30 minutes

Servings: 4

Ingredients:

- 1 tablespoon of vanilla
- 2 tablespoons of yogurt
- 6 limes
- 1 tablespoon of honey
- 3 eggs
- 6 oz of sugar
- 1 lb of soft cheese
- 2 oz of butter
- 1 cup of digestive biscuits

Directions:

1. Preheat the Air Fryer to 380F.
2. Cut the biscuits in the bowl.
3. Mix sugar and cheese well.
4. Add honey, 3 beaten eggs and vanilla to sugar and cheese.
5. Add juice of 6 limes and mix everything with yogurt.
6. Cook up to 15 minutes at 380F.
7. Then change the temperature to 350F and cook for 10-15 minutes.
8. Serve hot and after that enjoy with the delicious cheesecake with your family.
9. Prepare tea, juice or coffee with it.

Nutrition:

- Calories: 412
- Fat: 11,3g
- Carbohydrates: 27,9g
- Protein: 22,2g

Blueberry Cheesecake

Do you want to prepare something for summer? Then this cheesecake will be the great choice, because it is very delicious and has a lot of vitamins.

Prep time: 20 minutes

Cooking time: 25 minutes

Servings: 4

Ingredients:

- 5 tablespoons of icing sugar
- 1 tablespoon of vanilla
- 2 tablespoons of yogurt
- 3 oz of fresh blueberries
- 4 eggs
- 3 cups of sugar
- 1,5 lb of soft cheese
- 6 digestives biscuits
- 2 oz of butter

Directions:

1. Preheat the Air Fryer to 380F.
2. Chop the biscuits in the pieces and mix with butter.
3. Mix cheese and sugar in the bowl.
4. Beat eggs and add them to cheese.
5. Put there vanilla and yogurt.
6. Take the blueberries and mix them with all ingredients.
7. Cook cheesecake in the Air Fryer for 15 minutes at 370F and after that at 350 for 10 minutes.
8. Serve hot with tea.

Nutrition:

- Calories: 446
- Fat: 14,3g
- Carbohydrates: 24,9g
- Protein: 22,7g

Cheesecake with Caramel

If you like the sweets, you will not be able to refuse from this cheesecake. He incredibly delicious and not too complicated in the preparing.

Prep time: 20 minutes

Cooking time: 30 minutes

Servings: 4

Ingredients:

- 1 tablespoon of melted chocolate
- 1 tablespoon of vanilla
- 4 big eggs
- 7 oz of sugar
- 1 lb of soft cheese
- 6 digestives biscuits
- ½ cup of caramel

Directions:

1. Preheat the Air Fryer to 380F.
2. Then crushed the biscuits in the bowl.
3. Beat eggs and mix with the biscuits.
4. Then mix cheese, sugar, vanilla in the bowl.
5. Put cheese and rest of ingredients in the same bowl and mix everything well.
6. After that you can add the caramel.
7. Put everything in the Air Fryer.
8. Cook for 15 minutes at 380F, then 10 minutes at 370F and the last 5 minutes at 350F.
9. Take cheesecake out of Air Fryer and leave for 5-10 minutes.
10. At that time melt chocolate and cover cheesecake with it.
11. Enjoy with the delicious cheesecake. Do not forget about juice!

Nutrition:

- Calories: 458
- Fat: 16,3g
- Carbohydrates: 27,9g
- Protein: 25,7g

Cheesecake for Birthday

Do you have birthday and do not know what to prepare for your guests? The answer is very simple – the delicious cheesecake.

Prep time: 20 minutes

Cooking time: 30 minutes

Servings: 4

Ingredients:

- 1 melted chocolate
- 1 tablespoon of vanilla
- 2 tablespoons of honey
- 6 eggs
- 4 tablespoons of cacao powder
- 1 lb of sugar
- 2 lbs of soft cheese
- 2 oz of butter
- 6 digestives biscuits

Directions:

1. Break the biscuits into the small pieces and mix them with butter.
2. Put the bowl and mix the soft cheese with sugar.
3. Then add 5 eggs, honey and vanilla.
4. Mix everything well.
5. Beat the last egg and mix it with the cacao powder.
6. Cover cheesecake with it and put in the Air Fryer.
7. Cook 20 minutes at 380F, then 10 minutes at 370F.
8. Put it on the plate and cover with the melted chocolate.
9. You can server it hot or put in the fridge and server cold.
10. Your guests will be delighted with this cheesecake.

Nutrition:

- Calories: 449
- Fat: 16,1g
- Carbohydrates: 26,9g
- Protein: 25,4g

Conclusion

It seems that the Air Fryer is the technology of new generation. Today, the technologies are constantly developed and Air Fryer can improve your life completely. It is possible to prepare a lot of different meals in the Air Fryer and you will be surprised that it can help you a lot. Also, the cooked food cannot be combined with the food that is prepared in the common way.

Usually, you spend a lot of time to prepare some meal and if you have a big family or you invited a lot of guests, you can spend even the whole day at the kitchen, preparing the food. The recipes for the Air Fryer are unique and the process of the preparing food does not take the vitamins and minerals from products. The food is healthy and tasty.

You can find a lot of different recipes in this book and choose the recipes which you like. If you prepare them, you will see how it is possible to prepare the meal and do not spend a lot of time. You can start from the simple recipe and after that, step by step, prepare the most complicated. All recipes here are divided into the parts, like "Chicken", "Desserts" and a lot of other categories. You can choose any you like and cook.

It is possible to prepare almost all meals in the Air Fryer. It does not matter if you wish to prepare chicken or potatoes, desserts or vegetables. The main advantage of the Air Fryer is the fact that it saves a lot of your time. At that time, you can make everything very quickly and you will have time for yourself. Only a few minutes of your time and you can prepare simple and delicious meal. Your family will appreciate your efforts and you will be delighted with the result. You will have more free time and it means that the fact that you should prepare something for your family will create only the smile on your face because it is easy, simple, quickly and very delicious if you prepare the food in the Air Fryer.

As you can see, the Air Fryer has a lot of advantages, which can make your life easier:

- Multitasking

You can prepare any products, combine them and create a lot of new own recipes. You can try to prepare the food, which you have never eaten before and you will see that there is nothing complicated. If you worry that you can do something wrong, you can choose one of the recipe from the book, check it in detail and follow step by step instructions. You can be sure that the result will exceed all your expectations. You can find a lot of healthy meal recipes and change your life completely.

- The speed

All people know that the process of the preparing food is not very quick. However, if you have Air Fryer, it is not the problem for you anymore. You can prepare the meals even in 10-20 minutes. Do you think that it is impossible? Just try and you will see that you should not spend a lot of your free time on cooking because the Air Fryer can do everything for you. Just prepare the products, put them in the Air Fryer and set up the needed temperature. At this time, while you are doing something else, the Air Fryer will prepare the delicious meal for you and for your family. There is not any need for you to control the process, just wait and enjoy with the delicious and healthy meal.

- The start of process.

You can turn on the Air Fryer, put the products and even go to work while the food is preparing. When the food is ready, the Air Fryer will be automatically turned off. So, when you come, you will get the warm meal. It is very convenient. The Air Fryer is the great thing for people, which are too busy. The Air Fryer does not need that you should be near it all time. You will spend not too much time on the kitchen if you have Air Fryer. Your function will be only turn it on and off.

- The diversity

Some of the models of Air Fryer can prepare not only meals, but also bread, yogurt, some kinds of desserts and so on. You can choose any model of Air Fryer, choose the price and so on. It depends on your goals and needs. The Air Fryer is the great chance to change your life.

- The price

Today, there are a lot of different Air Fryers and you can choose the best price for you. These products are very popular and you can purchase them in almost every shop. You can choose the color, the price, the company which sells it and so on. This product will give you a lot of opportunities and if you purchase it one time, you will have it for many years.

- The safety

If you follow the instructions with the attention, you will be never sprinkled with the boiled oil from the Air Fryer and so on. This Air Fryer is very safe and you should not worry that your children can harm themselves. Because of this fact, it is the great advantage of the Air Fryer, which is important for everyone. Also, it does not take a lot of place on the kitchen. After using, you can wash the Air Fryer and put back in the box. As you can see, it is very convenient.

- The vitamins

If you prepare vegetables and fruits, a lot of vitamins are gone before the meal are ready. However, it is not connected with the Air Fryer. All vitamins and minerals are in the products when they are prepared in the Air Fryer. Because of it, you will get healthy and delicious food. It is very important for your body and you will see that the result from preparing the food in the Air Fryer will be great.

BONUS

Thank you for buying my book! I hope you enjoy it! Please follow this link below and download your FREE GIFT 220 Recipes Cookbook on home cooking!

https://goo.gl/1YKMR9

Made in the USA
San Bernardino, CA
05 December 2017